Vandervilson Alves Carneiro

Fieldwork in Geography

Vandervilson Alves Carneiro

Fieldwork in Geography

The conceptions of teachers and students

ScienciaScripts

Imprint
Any brand names and product names mentioned in this book are subject to trademark, brand or patent protection and are trademarks or registered trademarks of their respective holders. The use of brand names, product names, common names, trade names, product descriptions etc. even without a particular marking in this work is in no way to be construed to mean that such names may be regarded as unrestricted in respect of trademark and brand protection legislation and could thus be used by anyone.

Cover image: www.ingimage.com

This book is a translation from the original published under ISBN 978-613-9-72533-5.

Publisher:
Sciencia Scripts
is a trademark of
Dodo Books Indian Ocean Ltd. and OmniScriptum S.R.L publishing group

120 High Road, East Finchley, London, N2 9ED, United Kingdom
Str. Armeneasca 28/1, office 1, Chisinau MD-2012, Republic of Moldova, Europe
Printed at: see last page
ISBN: 978-620-7-92314-4

To Hebe, my wife, and my sons Rafael and Gabriel for their encouragement, patience, support and love at all times.

THE LIFE OF A TRAVELLER

My life is walking
Across this country
To see if one day
Happy rest
Saving the memories
The lands I've travelled through
Walking through the backlands
And the friends I left behind.
Rain and sun
Dust and coal
Far from home
I follow the script
One more season
And joy in the heart.
My life is walking...
Sea and land
Winter and summer
Show your smile
Show the joy
But not me
And longing in the heart
My life is walking...

Composition: Luiz Gonzaga and Hervê Cordovil

ACKNOWLEDGEMENTS

I would like to say a huge THANK YOU to the people who have contributed directly or indirectly to the realisation of this work:

- To Professor and friend Manoel Rodrigues Chaves for his guidance in this work.
- To my sons Rafael and Gabriel, my wife Hebe and also to my parents Maria Luiza and Euclides and my brothers Leandro, Sandro, Nenê and Sérgio for their companionship, friendship, affection and understanding in tense moments.
- I would like to thank Professors Eguimar Felício Chaveiro and Francisco Carlos de Francisco for agreeing to take part in my master's programme and for the geographical and professional interaction.
- To my work colleagues and students on the Geography course at the State University of Goiás, in Pires do Rio, for their support and companionship.
- To my work colleagues and students on the Geography course at the Federal University of Goiás in Catalão, for their support and companionship.
- Professor and friend Adriany Ávila de Melo Sampaio, from the Geography course at the Federal University of Uberlândia, for her encouragement, support, friendship and geographical and professional interaction.
- To my friends Reginaldo dos Santos and Reginaldo Moreira, professors at the Faculdade do Sudeste Goiano, for their technical support and statistical suggestions.
- To my friend Marcelo Venâncio, a professor at UEG - Pires do Rio, for reading, contributing to and technically formatting this work.
- To my friend Sônia Martins, a professor at UEG - Pires do Rio and UFG - Catalão, for reading, contributing, writing the abstract and correcting the work through the lens of the Portuguese language.
- To my friends and fellow master's students Sebastião, Tiago Godoi, Lucineide, Vandério, Mirne, Elaine Lobo, Kharen for the coffees, sweets, academic work, meals, events, trips, pure friendship, conviviality and encouragement.
- I would also like to say hello to the rest of my colleagues on the master's programme and happy geographical journeys.
- To Professor and friend Gustavo Adolfo Rocha from UEG - Palmeiras de Goiás, for his support, loan of material, contributions and professional interaction.
- Professor and friend Valeriê Machado from CEFET - Morrinhos, for her sincere friendship, support, contributions and socialising at geographic events.
- To friends and teachers Sérgio Pereira de Souza (São Paulo State Public Education Network / Presidente Prudente Municipality) and Zilda de Fátima Mariano (Federal University of

Goiás / Jataí Campus) for their geographical energy since graduating in Geography at Presidente Prudente / UNESP, loyal friendship, contributions and exchange of experiences.

- To my fellow teachers at the House of Teachers at UEG - Pires do Rio, for the exchange of experiences and the tenuous relationship between the peaceful and the explosive.
- To the staff of NEPEG - Centre for Studies and Research in Geography Teaching, for the professional interaction.
- To UEG - State University of Goiás for the partial licence to carry out my master's research.
- To Casa da Sopa Fabiano de Cristo (Senador Canedo / GO), for their spiritist studies and voluntary work with the needy children of the São Sebastião neighbourhood.
- To our spiritual friends from Uberaba / Minas Gerais, Dr Odilon Fernandes and Dr Inácio Ferreira, for their daily encouragement and support in our work.

SUMMARY

Fieldwork is one of geography's most traditional didactic resources for the pursuit of knowledge. This resource has been a feature of both university and primary and secondary school environments. This research aims to find out about the conceptions of teachers and students about the didactic resource fieldwork in Geography Degree Courses in the south-east of Goiás. The data was collected through questionnaires with teachers and students in their study and work environments. Analysing the results showed the conceptions of fieldwork held by both teachers and students. A more detailed investigation using the graphs showed that there are significant conceptions among teachers and students regarding the didactic resource called fieldwork, applied to the subjects of the courses in question. Given the various fieldwork alternatives, the students and teachers at the State University of Goiás, Pires do Rio University Unit opted for the motivating type modality and the teachers and students at the Federal University of Goiás, Catalão Campus chose the training type modality. Regardless of the choice of type and/or modality of fieldwork, teachers and students must be clear that this activity is very important for geographical knowledge and it is through it that landscapes, places and municipalities are unveiled.

Keywords: Fieldwork. Geography degrees. Teaching resource. Students' and teachers' conceptions.

SUMMARY

INTRODUCTION	**7**
CHAPTER 1	**18**
CHAPTER 2	**80**
CHAPTER 3	**136**
CHAPTER 4	**196**
CHAPTER 5	**198**
CHAPTER 6	**213**

INTRODUCTION

I lived with my maternal grandparents for a while until I was eight years old, in the countryside of Guarulhos / São Paulo (Capelinha region), where I discovered the slopes, springs, ponds, marshes, streams and rural properties of the Serra da Cantareira. Here, the geographical landscapes "sprouted" with each new discovery.

School age had arrived and so I went to live with my parents on the outskirts of Guarulhos (in the metropolitan area of São Paulo, in the neighbourhood of Cumbica), to start and finish my initial studies (the former 1st and 2nd grades).

In the old 1st and 2nd grades, we never did any fieldwork. Whenever possible, the school took us to the Playcenter amusement park in the capital of São Paulo, along the Avenida Marginal do Rio Tietê. The rides caught my attention, but not as much as the traffic jams on the avenues, the gigantic buildings, the industries, the favelas, the heavy rain and the stench of the River Tietê.

Even so, as a curious, studious boy, a ball player and a bit of a wanderer, I would wander around the neighbourhood of Cumbica and the surrounding area. My walks always showed things that the geography textbook didn't cover.

The job came when I was around fourteen years old in a chocolate industry, where I was able to get to know almost all the large municipalities in the São Paulo Metropolitan Region in more detail through my job as an office boy.

From then on, I worked at paper companies and airport services until I took the entrance exam.

In my final year of secondary school, I decided to take the entrance exam for a degree in Geography, as I had been influenced by my teachers since primary school.

I applied for the entrance exam at VUNESP - Fundação Vestibular da Universidade Estadual Paulista, and was successful in my application to study Geography.

When I filled in the application form, my aim was to study Geography at UNESP in Rio Claro, as it was closer to Guarulhos (about 200 kilometres). However, an oversight when filling in the application form, a code error, led me to study Geography at UNESP in Presidente Prudente, in the heart of western São Paulo, the cradle of the struggle for agrarian reform in São Paulo. This error in filling in the code took me some 550 kilometres further away from Guarulhos.

In life, mistakes happen and they are lessons learnt. In Presidente Prudente, I was able to broaden my horizons, live in student unions, get to know the neighbouring municipalities, get to know Geography for real, because the Geography in the textbook was very different.

In this way, I was able to realise what Ab'Saber (2001, p. 15) argued, "an efficient education is only possible when the student is intimate with the place where they live". This was the case in Guarulhos and Presidente Prudente.

Little by little, I became involved in the student movement; I carried out voluntary monitoring activities in the Geology Laboratory and also assimilated Troppmair's (1988, p. 13-16) words about the scientific plan, i.e. the research plan:

> 1- Research is first and foremost a work of patience; 2- Choose a topic that piques your interest; 3- Clearly see the problem you want to research; 4- Draw up the plan yourself; 5- Think logically and coherently; 6- A plan is always preliminary; 7- Spend time drawing up the plan; 8- Use correct and precise language; 9- Writing a scientific paper always requires time.

It should also be said that according to Gil (1999, p. 19),

> Research can be defined as a rational and systematic procedure that aims to provide answers to problems that are proposed. Research is required when there is not enough information available to answer the problem, or when the available information is in such disarray that it cannot be adequately related to the problem.
>
> Research is carried out using available knowledge and the careful use of methods, techniques and other scientific procedures. In reality, research develops over the course of a process that involves numerous phases, from the appropriate formulation of the problem to the satisfactory presentation of the results.

He also says that the success of a research project depends fundamentally on certain intellectual and social qualities of the researcher, among which we highlight: "a) knowledge of the subject to be researched; b) curiosity; c) creativity; d) intellectual integrity; e) self-corrective attitude; f) social sensitivity; g) disciplined imagination; h) perseverance and patience; i) trust in experience" (GIL, 1999, p. 20).

On holidays and long holidays, I would go home to Guarulhos by train, bus or hitchhiker. I remember the landscapes of the interior of São Paulo unfolding before my eyes. The windows of the train, bus and hitchhiker's car acted as a viewpoint for observing the geographical mosaic of the landscape.

Fieldwork has always been part of my daily university life since 1990, when I was studying for my degree in Geography at UNESP - the University of São Paulo.

Universidade Estadual Paulista "Júlio de Mesquita Filho", in the municipality of Presidente Prudente / State of São Paulo.

I say that before university, my wanderings and observations were unpretentious and amateurish, but memorable.

This university enabled me to get to know the Central-Southern region of Brazil through fieldwork, under the coordination of Prof Dr João Lima Sanfanna Neto (Climatology), Prof Dr Eliseu Savério Sposito (Evolution of Geographical Thought) and Prof Dr Roberto Braga (Regional Planning). Roberto Braga (Regional Planning), and also my involvement in a scientific initiation project on geographical studies of the Mato Grosso Amazon, coordinated by Prof Dr Messias Modesto dos Passos (Geography of Waters: Continental and Oceanic).

I would also like to highlight the great contribution of fieldwork in the subjects of

Geology, taught by Prof Dr Manoel Carlos Toledo Franco de Godoy, Geomorphology, taught by Prof Dr Hideo Sudo and Biogeography, taught by Prof Dr Francisco Carlos de Francisco.

After 1993, in the capital of São Paulo, I began my professional career in Geography, teaching in private supplementary schools (formerly 1st and 2nd grades) and preparatory courses for competitive examinations. In these schools, in the centre and south of the city of São Paulo, I always used fieldwork to get to know and understand the contrasts of the great metropolis of São Paulo.

In support of this, I see that according to Ab'Saber (2001, p. 15),

> That's where geography comes in, with its ability to help students understand the place where they live. Only then will they be able to act on that environment later on. That's why every teacher needs to understand their surroundings, their population and their problems. It's not enough to know the basics and use them in inconsequential readings of old textbooks.

Later, in 1996, I was invited to work at FAESCI - Celso Inocêncio de Oliveira State College (now UEG - Goiás State University), in Pires do Rio. There, I began teaching Physical Geography I (Geomorphology and Climatology), Physical Geography II (Hydrography and Soils) and Systematic and Thematic Cartography.

In Pires do Rio, I was able to put fieldwork into practice in an attempt to get to know the place where the students live better. The work flowed, and little by little I was able to draw up a project for the creation of the NAC - Núcleo de Aulas de Campo - and put it into practice from 1996 until 1999.

In 2000, UEG took over FAESCI and the NAC - Núcleo de Aulas de Campo (Field Lessons Centre).

was renamed NUGAC - Núcleo Geográfico de Aulas de Campo. With the name NUGAC, the next step was to have an electronic magazine to publicise academic work. This goal was achieved in 2006 with Revista Mirante. This publication is still available to the geographical community.

I realise that fieldwork has always been with me on my geographical journeys, which is why this research is justified with the aim of gaining a deeper understanding of the subject of fieldwork in Geographical Science.

According to Corrêa (1999, p. 4),

> The complexity of spatial organisation on the threshold of the 21st century, in a globalised world, is an issue for geographers to consider and deal with in their work. This involves both the need to rethink the theories that describe and explain the ever more rapidly changing spatial organisation, as well as the set of procedures aimed at obtaining and processing pertinent and relevant information to make the reality seen geographically intelligible. These include fieldwork.

This master's research aims to find out about the conceptions of fieldwork developed in the Geography Degree programmes in the south-east of the state of Goiás, where UEG - State University of Goiás, Pires do Rio Campus and UFG - Federal University of Goiás, Catalão Campus are located.

Fieldwork in Geographical Science is an essential tool and deserves to be highlighted so that it doesn't fall into the trivialisation of mere tours and tourist excursions.

According to Kayser (1985, p. 34-35),

> The researcher must be careful not to be distracted by the anecdotal, the strange, the singular.
> It's one thing to observe in order to try to understand, to record phenomena in order to interpret them with the support of a general explanation; it's quite another to go "researching" like someone who goes to the zoo or on safari.

It should be noted that:

> Geography is an area of knowledge that has been valued since antiquity. Maintained and studied by the Arabs during the Middle Ages, it contributed to the vast domination of their civilisation. In Europe, with the Renaissance in the 15th century, it collaborated with the expansionism represented by the great navigations and colonial conquests. Hence the emergence of Geographical Societies in various countries, bringing together religious people, merchants, navigators, naturalists, politicians, military men, philosophers, etc. These societies functioned as a privileged locus for the discussion of geography until the 20th century. With the encouragement of these institutions, Geography established itself as a school subject, becoming basic to people's general education (PINHEIRO, 2006, p. 91).

Thus, ever since the travellers and naturalists (Humboldt, Pohl, Saint-Hilaire and others) and the great schools of geography, fieldwork has always been present, fulfilling

certain purposes, such as administrative, military, territorial knowledge, use and occupation of places, economic, among others.

In this vein, Professor Gomes (1999, p. 13) states that "naturalists have bequeathed us a valuable collection of documents, the fruit of their expeditions, journeys and excursions through the lands of Goiás. Examples include Pohl, Saint-Hilaire, Castelneau, etc". These were the same naturalists who unveiled Brazil through fieldwork.

According to Fonseca and Kuvasney (2003, not paginated),

> For a long time, travelling was the main procedure for expanding knowledge about the earth's surface and other geographical spaces. Those who undertook these journeys came to control knowledge whose use served both economic interests and projects of conquest and colonisation. To put it another way: travelling represented an increase in the geographical scale of human relations, with European peoples at the forefront as far as the Western world was concerned.
> In the 15th century, a journey from Paris to Rome could be an adventure of a lifetime. People's activities were limited to a radius of 5 to 12 kilometres around their home. Setting out on long journeys was for the few. In that sense, the world was small. Europe ran from the Atlantic to Kiev; to the south the world was "closed", with Constantinople being the last port.
> Today, on the other hand, we have a view of the whole world, even if we don't know it in situ. Our borders are surpassed by an impressive expansion of our gaze, expanded by satellite images, which give us the whole world and portray the geometry of objects, "not really geographies, because they come to us as objects in themselves, without society living inside them". In order to know the true value and meaning of things, we need to add other means to the technological one. One very interesting resource is fieldwork.

In Brazil, exactly from the 1930s to the 1950s, under the influence of Geography

The French Geography Schools of Pierre Defontainnes, Pierre Monbeig and Francis Ruellan appeared in São Paulo (1934) and Rio de Janeiro (1935).

Fonseca and Kuvasney (2003, not paginated) say that "the importance of fieldwork in Geography has been known to us since Herodotus and Strabo".

However, Monbeig (1936 apud Fonseca and Kuvasney, 2003, not paginated) argued that "excursions are a valuable aid and should be used and applied with a defined, geographical objective, so that they do not become a mere stroll or tourist trip".

Professor Ab'Saber (2001, p. 16), recalling his first scientific excursion, says:

> On the first day of class, Professor Pierre Monbeig organised a field trip. We left São Paulo for Itu, Salto, Campinas and Jundiaí. Until then, my geographical knowledge had been limited to São Luiz do Paraitinga and the surrounding area. Come to think of it, that wasn't my first memorable trip. When I was five, my father took us to Ubatuba. We went on horseback along the old Estrada do Café, which was abandoned. I carried a jacá (a basket used to carry food on the backs of animals) on one side and my younger brothers on the other. We passed the farms that surrounded the town, entered the transition zone, with subsistence agricultural production, and passed through private but unused land. On the trail, we got to know the forest that precedes the Serra do Mar. There was a lot of water dripping from the leaves, as this is a humid region, like all high mountain areas. When I went on the college excursion, I felt like it was the continuation of an interest that had sprung up on that trip to Ubatuba.

Over the years, from the 1960s to the 1980s, the core of fieldwork, i.e. the observation and description of places, was seriously criticised by Marxist and quantitative geographers. Marxist geographers claimed that fieldwork didn't address socio-economic issues and was therefore a positivist and inadequate practice. And quantitative geographers valued statistical-mathematical data to understand reality, leaving fieldwork aside.

From 1990 onwards, fieldwork received new impetus as a means of getting to know places. Because of the financial conditions of students and the lack of infrastructure (buses) on the part of public higher education institutions and public middle and elementary schools, the option was made to carry out this activity in the municipality and surrounding areas where the schools are located.

Silva (2006, p. 52), makes it clear that "in Geography we can still count on other resources that tend to raise the quality of research, namely fieldwork".

From this point of view, Lima and Assis (2004/2005, p. 112) emphasise that "the Fieldwork is configured as a resource for students to understand the place and the world, linking theory to practice, through observation and analysis of the lived and conceived space".

Thus, through informal conversations with teachers and students at scientific geography events across the country, we have found that there are different conceptions of the fieldwork tool in public higher education.

In view of the above, we are concerned about the way these field activities are conducted and developed. That's why I have a few questions:

1. What is the importance of fieldwork in Geography degrees in the Southeast? Goiano?
2. What are the conceptions of fieldwork among teachers and students?
3. What methodologies have been developed for fieldwork?
4. What are the contributions of the naturalists and the geographical schools to labour? in the consolidation of Brazilian Geography?

We believe that fieldwork is a modality that allows us to reflect on the place, discuss the articulation between theory and practice and underpin teaching and learning.

Mao Zedong (1941 apud Seleção de Textos, 1985, p. 42), states that:

> There are many who, "barely out of their prams", boast, making speeches, distributing their points of view, criticising this and criticising that; in fact, out of every ten of these people, ten know failure. Because their speeches, their criticisms, are not based on any detailed fieldwork, they are nothing more than talkers.

He also emphasised that "for those who only understand theory without knowing anything about the real situation, carrying out such field research is even more necessary, otherwise they will not be able to link theory to practice" (Mao Zedong, 1941 apud Seleção de Textos, 1985, p. 42).

The practice of fieldwork means valuing the discovery of the place where it is not carried out at random, but rather clarifies, fixes, sees, reviews and experiences the topics covered by the subjects in the classroom and fosters debates and interventions in the socio-environmental reality.

It is therefore necessary to go into the field, because contact with reality in itself determines the start of an entire learning process.

In fieldwork, there is something to see and think about in geography. It allows students to analyse space, which is an essential geographical task. What we need is to know how to look, how to describe and make correlations in the space in question.

According to Kayser,

> If fieldwork is not to be just empiricism, it must be linked to theoretical training, which is also indispensable. Knowing how to think about space doesn't just mean placing problems in a local context; it also means effectively linking them to phenomena that develop over much wider areas (1985, p. 20).

The great contribution of this research will be to show students and teachers of Geography degrees in south-eastern Goiás and elsewhere in Brazil that there are other approaches, other views on fieldwork, how to do it, how to avoid doing it for the sake of doing it, how to say that the municipality has something to study, see, reflect on, discuss and also how to use coherent and effective methodologies to establish a good teaching-learning process.

Of course, for the geographer, field practice means a laboratory where he can clearly see and size up all the theoretical, physical and human aspects learnt in the classroom or office.

Santos says:

> The importance of fieldwork is not just limited to listening to people; to the meaning they give to things; to what is finished; nor to reality as a realised and finished fact. The importance of the empirical, therefore, is to promote contact, in other words, it is the analysis focused on the tendencies of interpretations that researchers promote of the world in a dynamic movement orientated by the social determinations of their place. This procedure implies an understanding of the lived experience, which derives from the practical acts that people, based on their social organisations, construct in time and space (1999, p. 120).

Exploring the physical and social environments through fieldwork is a fundamental tool for this reading of educational practice.

According to Moço (2008, p. 71), "Geography teaching should be based on field trips, reading texts of all kinds and producing and interpreting maps".

Professor Lacoste goes on to say that:

> At the moment there are certainly too few credits, but you don't have to go far to find something to observe and research. The ground can initially be in the vicinity of the university: walking is the main way to get around. It's not so much a question of going on excursions or organising internships to learn this or that technique (1985, p. 15).

Fieldwork develops the subject's ability to operate, execute, compare, explain, debate, analyse and draw conclusions. The teacher should not be anchored only in the accumulation of geographical knowledge from the textbook, but should move away from exhaustive speeches, questionnaires without a foundation, intensify communication with students, update and improve knowledge and be happy to experiment with new techniques (TOMITA, 1999, p. 13-15; SECRETARIA MUNICIPAL DE EDUCAÇÃO DE SÃO PAULO, 1992, p. 40-43).

According to Waibel (1979, p. 34), "by observation we do not mean simply seeing, the autopsy of the landscape, because observing is the act of seeing linked to the act of thinking, and it means that one has to interpret what is seen".

He goes on to say,

> I've often been asked why I started my fieldwork in the state of Goiás, which is so remote and "wild". The answer is simple: as I was interested in colonisation, I had to go inland; as I was also interested in learning about the original vegetation and its transformation by human activity, I decided to go to a region where human influence was kept to a minimum. I would have preferred to start my work in Mato Grosso. But as the central part of this state is not accessible by railway, I had to go to Goiás, where the railway penetrates further inland. I think my observations of the vegetation in the south of Goiás will prove that it was a good idea to start my work in Brazil from the rear, so to speak (WAIBEL, 1979, p. 160).

It's worth saying at this point that fieldwork isn't everything and it's not just a tool for Geography either. Other branches of knowledge also use this didactic resource, such as Anthropology, Geology, Biology, among others.

Coltrinari (1996, not paginated) argues that:

> Fieldwork is not an invention of geographers, nor are they the only ones who practice it; inherited from naturalists and earth scientists by their successors who are now trapped in the networks of the division of scientific labour, it has become, in some areas of knowledge, an irreplaceable source of information.

Marandola Jr. and Lima (2003, p. 175) state that:

> The practice of fieldwork is not a methodology exclusive to Geography, but is also widely used by the other sciences. Due to the difference in approach and perspective, work in the social sciences and the physical (natural) sciences is often referred to as having variations in methodology and orientation. However, in both cases, fieldwork is seen as crucial to the development of knowledge.

I assert that geographers also need to be aware of the fundamental importance of fieldwork as a tool for knowledge and research (LACOSTE, 1985, p. 15).

For this to happen, teachers need to get out of their encampment, their fiefdom, their partisanship, their exaggerated/extremist militancy and do something new in favour of scientific knowledge, because, according to Santos (2000, p. 7),

> I can even be a militant from time to time. However, the great nourishment of the intellectual is free thought and that's why he can't be chained to parties, militancies or groups. The intellectual has to be a militant when the country is in danger, which is the current case in Brazil. Then you can join the military. But if you've been a militant all along, you could miss the opportunity for a breakthrough. Militancy is what happens inside universities, which are factories of mediocre people. Universities are controlled, we have closed groups within universities and these groups create mediocre people or facilitate the production of mediocre people. Great opportunities arise for those who belong to a group, but from the point of view of great intellectual production, you fall behind. This isn't the only problem universities have, they have several problems, which they don't look at or don't want to look at because of their perhaps crystallised practices.

The general objective of this master's research is: TO KNOW THE CONCEPTIONS OF TEACHERS AND STUDENTS ABOUT FIELD WORK IN GEOGRAPHY LICENCES IN SOUTHEASTERN GOIANO and the specific objectives are: a) verify the importance of fieldwork for higher education in Geography, b) identify the conceptions of teachers and students about fieldwork, c) score the methodologies developed about fieldwork in Southeastern Goiás, d) compare the methodologies developed about fieldwork in Southeastern Goiás and e) detect the contributions of naturalists and geographical schools about fieldwork to the constitution of Brazilian Geography.

The starting point for the research methodology is the theoretical background. To begin with, a bibliographical survey was carried out with readings and reflections on the subject of FIELDWORK in Geographical Science.

According to Gil,

> Bibliographical research is carried out using material that has already been prepared, consisting mainly of books and scientific articles. Although almost all studies require some kind of work of this nature, some research is carried out exclusively from bibliographic sources. A large part of exploratory studies can be defined as bibliographical research. Research into ideologies, as well as research that aims to analyse the various positions on a problem, are also usually developed almost exclusively from bibliographic sources (1999, p. 48).

The methodological procedures were then organised: the choice of higher education institutions with degree courses in Geography in south-eastern Goiás (State University of Goiás - UEG, Pires do Rio Campus and the Federal University of Goiás - UFG, Catalão Campus). Taking a brief look at the research area, the cities of Pires do Rio and Catalão are located in south-eastern Goiás, in what is known as the Railway Region, and have distinct socio-economic characteristics, as explained in Chapter 2 of this dissertation.

Two types of questionnaire were then drawn up: one for the teacher and one for the student, to gauge their conceptions of fieldwork. We also used Arabic numerals to differentiate between the same questionnaires and those involved; we also chose to preserve the identification of those surveyed.

Professor Gil says that "a questionnaire is a set of questions that are answered in writing by the respondent" and "the questionnaire is the quickest and cheapest way of obtaining information, as well as not requiring staff training and guaranteeing anonymity" (1999, p. 90).

It is worth remembering that the Geography degree courses at both institutions are offered in the evening and the questionnaires were administered to the 1st, 2nd, 3rd and 4th year classes. From November 2007 to May 2008, the questionnaire was administered at UFG to 132 students and 19 teachers, and at UEG to 142 students and 13 teachers. These questionnaires identify the profiles of the two institutions, the students and the teachers; the role, importance and negative and positive aspects of Fieldwork in Higher Education from the point of view of the students and teachers.With regard to the one hundred and forty-two (142) questionnaires applied to the students of UEG - Pires do Rio, we see that 54% were returned and 46% were not returned (Graph 1).

Graph 1: Questionnaires applied to students at UEG - Pires do Rio
Source: Research carried out from November 2007 to May 2008
Org.: CARNEIRO, V. A. - July/2008

At UFG - Catalão, of the one hundred and thirty-two (132) questionnaires applied to to the students, we found that 47 per cent returned it and 53 per cent didn't (Graph 2).

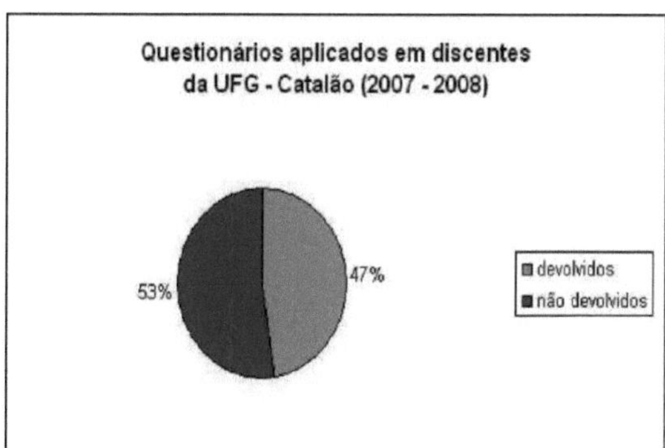

Graph 2: Questionnaires applied to students at UFG - Catalão Source: Research carried out from November 2007 to

May 2008
Org.: CARNEIRO, V. A. - July/2008

In Catalão, the questionnaire was administered to nineteen (19) teachers, 47 per cent of whom returned it and 53 per cent of whom didn't (Graph 3), and in Pires do Rio, to thirteen (13) teachers, 85 per cent of whom returned it and 15 per cent of whom didn't (Graph 4).

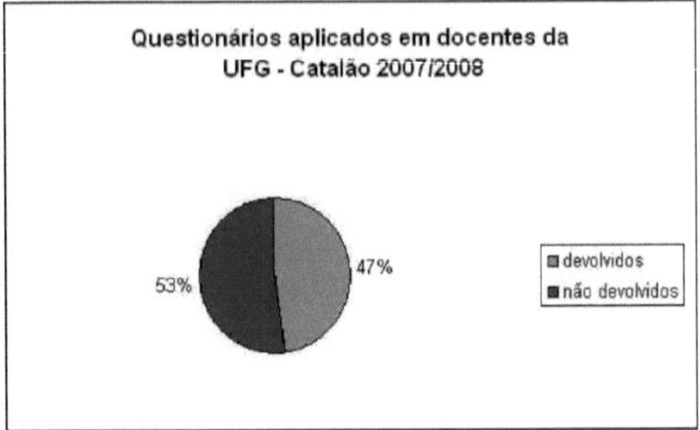

Graph 3: Questionnaires applied to teachers at UFG - Catalão Source: Research carried out from November 2007 to May 2008
Org.: CARNEIRO, V. A. - July/2008

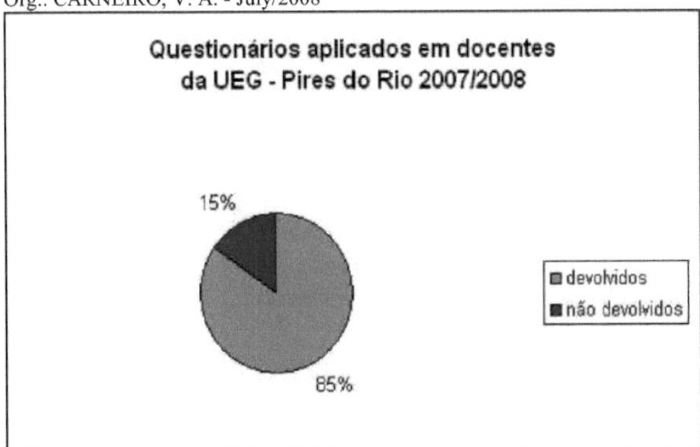

Graph 4: Questionnaires applied to teachers at UEG - Pires do Rio
Source: Research carried out from November 2007 to May 2008
Org.: CARNEIRO, V. A. - July/2008

I'd like to point out that when it comes to drawing up the graphs for teachers and students, we've opted to draw them up individually, i.e. we'll have graphs for teachers and students in Pires do Rio and also for Catalão. All the graphs were produced using the Microsoft Office Excel 2007 programme.

Another fundamental step was to carry out a documentary survey (minutes, diaries, regiments, ordinances, pedagogical political projects, laws, decrees, and others) at the two institutions in order to compile a brief history of the courses in question and detect the pedagogical conduct of the fieldwork.

For Gil,

> Documentary research is very similar to bibliographical research. The essential difference between the two lies in the nature of the sources. While bibliographical research makes fundamental use of the contributions of various authors on a given subject, documentary research makes use of materials that have not yet received an analytical treatment, or that can still be reworked according to the objects of the research. Documentary research follows the same steps as bibliographical research. The only thing to bear in mind is that while in bibliographical research the sources consist mainly of printed material located in libraries, in documentary research the sources are much more diverse and dispersed. On the one hand, there are "first-hand" documents, which have not been analysed in any way. This category includes documents kept in the archives of public bodies and private institutions, such as scientific associations, churches, trade unions, political parties, etc. This includes countless other documents such as personal letters, diaries, photographs, recordings, memos, regulations, letters, bulletins, etc.
> On the other hand, there are "second-hand" documents that have already been analysed in some way, such as research reports, company reports, statistical tables, etc (1999, p. 51).

It is important to make it clear what is meant by fieldwork. In this regard, I use the viewpoint of Castrogiovanni and Schutz (1986, p. 43), "we understand fieldwork to be any activity that takes place outside the classroom and seeks to realise stages of knowledge and/or develop skills in concrete situations through observation and participation".

The dissertation is divided into an introductory section, three (3) central chapters, final considerations and annexes.

The first part, the introductory part, presents an account of the reasons that led me to research the subject of fieldwork, using a brief memoir as a resource, as well as setting out the objectives and methodological steps.

In the second part, which contains the three (3) central chapters, we have:

Chapter 1 "FROM RUDIMENTS TO THE SYSTEMATISATION OF GEOGRAPHICAL SCIENCE" looked at the trajectory of geographical science from primitivism to the present day; it also focused on how fieldwork is developed within geographical schools and currents and, for the Brazilian case, how fieldwork has been used to unveil the territory from the colonising period to the present day.

Chapter 2 "CATALÃO AND PIRES DO RIO: DIFFERENT CONTEXTS IN THE GIOIANO CERRADO" highlights some of the physiography and the processes of use, occupation and degradation of the Cerrado where the two higher education institutions (UEG and UFG) are located; it also looks at some of the geo-historical background of Catalão and Pires do Rio that fostered the creation of these two higher education units and their Geography courses, as well as highlighting the emergence of higher education in the state of Goiás.

Chapter 3 "FIELD WORK: CONTRIBUTIONS AND CONCEPTIONS" highlights the great contribution of Humboldt and other researchers, showing the types of fieldwork, and finally presents and analyses the results of the teachers' and students' conceptions of fieldwork.

In the third part, we have the references (materials used for the master's research), the final considerations of the author of this dissertation and the most significant annexes.

CHAPTER 1

FROM THE RUDIMENTS TO THE SYSTEMATISATION OF GEOGRAPHICAL SCIENCE

This chapter will show the evolution of geographical knowledge on fieldwork from primitivism to the scientific organisation of Geographical Science, including the illustrious geographers Alexander Von Humboldt, Karl Ritter, Friedrich Ratzel and Paul Vidal de la Blache to the geo-Marxist current of Yves Lacoste and Pierre George. It will include an approach to Geography in the Brazilian scenario and it should also be remembered that the theme of fieldwork will be approached and verified within the biases of the geographical currents.

1.1. Some notes on the trajectory of Geographical Science

The following notes will provide a sufficient basis for understanding the lineage of fieldwork in Geographical Science during the academic and non-academic stages.

Thus, we can say that geographical knowledge dates back to the "primitive times of human history" (JUNQUEIRA, 2003, p. 48; RODRIGUES, 2008, p. 35), a geological period called the Pliocene. A competitive scenario, a search for food and knowledge of their surroundings.

In this line of thought, Adas (1982, not paginated) believes "that it is not possible to study Geography, and especially Human or Social Geography, without a historical basis". Thus, according to historians such as Aquino et al (1990, p. 19), Man appears "in the distant past, which corresponds to the end of the period that archaeologists and historians call the Palaeolithic Age (Age of Chipped Stone), and geologists call the Pleistocene".

For Guerra (1989, p. 341), the Pleistocene is a "period that follows the Pliocene and marks the beginning of the Quaternary" and "the first hominids are attributed to the Pliocene". In Magalhães Filho (1983, p. 1),

> From the geological period known as the Pliocene, which began around ten million years ago, the slow march of evolution led to the emergence of several species in the primate order that are characterised by their upright stature, lack of a tail and larger brains.
> These various species, grouped together for the sake of simplification as the family of hominids, were made up of animals that, in order to survive in the struggle for existence, had been forced to leave the forests in which their ancestors had lived.
>
> Their ancestors had to learn how to earn a living and defend themselves in savannahs and prairies, where the conditioning inherited from the previous type of life from which they came was of little value.
> Over millions of years, these related species went through a slow process of adaptation and selection. It was a question of creating new conditions suited to the actual living conditions they had to face. And the challenge was made even greater by the fact that the geographical environment itself was undergoing continuous changes as the next geological period approached, the Pleistocene, which would bring with it the great glaciations that would cover large swathes of the earth's surface with glaciers.
> In this marvellous process of responding to nature's challenge, some of the hominid species managed to develop skills capable of allowing them to survive and evolve, while others were condemned to extinction. Among the former, the continuous effort to obtain food, and to avoid becoming food, led to the use of objects taken from the very natural environment in which they lived.
> This use of objects picked up from the natural environment (stones, bones, pieces of wood) was certainly a mere reflex action at first. And as a reflex action, mere instinctive conditioning, it remained for generations. But the successive, constant repetition of this reflex act, complex by its very nature because it necessarily included an object external to the animal that practised it, further stimulated the development of the brain capacity of these species.
> It will never be known exactly when or where, and it may have happened in several places, at different times, and it was also a slow process in itself. What is certain is that one day the animal became aware of the reflex act it was practising. And it became aware of the distance between itself, the subject, and the object it was using. And it became aware of the relationship that existed between the two at the time of the act.
> Consciousness was born from labour; with consciousness, the reflex act became labour. Through labour, by generating consciousness, the animal was transformed into man.

Sodré (1976, p. 13) emphasises that:

> Geography is perhaps the science with the longest history. It begins, in fact, with the first gentile communities. One of the most curious facts that the study of prehistory reveals to us is certainly the tendency towards movement which, along with early human dispersal,

contributed to the first knowledge of regions other than those in which communities first lived and where they established relationships with nature.

We know very little about the great mobilisations (migrations) of the period before history, and the geographical obstacles, hydrography and relief, were no impediment to human dispersal, as there are traces of man everywhere. According to Sodré (1976, p. 13-14):

> Whatever the reasons for the earliest displacements and migrations, ranging from necessity to adventure, they led to a broader knowledge of the Earth's surface and the tendency to record or pass on this knowledge. This is undoubtedly geographical material. The fact that it is empirical is of little importance in this case. What matters is whether the knowledge can be transmitted and retained. Consequently, the recording of this knowledge is of particular importance. So, if we accept a division of the history of geographical events, not to mention geography, that includes the preliminary stage of prehistory, it would include a subdivision: before the record and after the record.

According to Andrade (1992, p. 20), "if we accept that primitive peoples were those who lived in prehistoric times, with no knowledge of writing, we are forced to admit that, living on the surface of the Earth, deriving their sustenance from it and having a conception of the world, they already had geographical ideas". Continuing with the same author,

> When we talk about primitive peoples, considering them as those who lived in prehistoric times, we see that even though they didn't have writing, passing on their knowledge through word of mouth and drawings on rocks and in caves, passed down from generation to generation, they had a concept of life and a culture, both impregnated with geographical ideas (1992, p. 20).

In Dollfus (1991, p. 29):

> Human action tends to transform the natural environment into a geographical environment, that is, an environment shaped by human intervention in the course of history. This is a recent development in the history of the world. In fact, although palaeontology tells us about the appearance of beings that we can consider to have been the first men, which took place two million years ago in East Africa, the role of man as an agent of intervention in geographical space only dates back 6,500 or 7,000 years, with the beginnings of agriculture. The generalisation of agriculture took place in various regions of the world three or four millennia ago. However, human activity has become increasingly intense, thanks to the combined effects of population growth around the world and technical progress. So this thin film that is human history in relation to the thickness of world history nevertheless occupies a place of capital importance for understanding and explaining geographical space.

Silva and Fioreze (1999, p. 251) emphasise that:

> It was in the midst of the organisation of primitive human societies that what would come to be known as the modes of production arising from the relationships they established between themselves, within communities and the relationships they established with nature, or simply the natural framework from which they suffered very direct influences, from which they would free themselves when they developed technically, was outlined.

Rodrigues (2008, p. 10) states that "over the centuries, man has sought to know the world around him and its objects. Human history has been the history of knowing and understanding nature. It is the history of appropriating and transforming nature".

Thus, the historical scenario is known for primitive economies whose character is established by nomadism, and later reach the stages of agricultural economies (agriculture and grazing of small animals) based on human sedentarism.

Man's attachment to the land is only possible through knowledge of the places that favour the choice of fertile soils, flowing bodies of water, the presence of biodiversity and natural fortresses for protection.

During the phase in which man was nomadic and became sedentary, historians were only concerned with the distribution of man over geographical space, known as historical geography. In the last decades of the 20th century and the beginning of the 21st century, however, they learnt to look at the magnitude of man's transforming relationship with the physical-environmental environment that surrounds him (FONTANA, 2000, p. 11).

With the passage of time, "from the primitive community to the collecting and producing communities" (AQUINO et al, 1990, p. 21), "from Geography before the record to Geography after the record" (SODRÉ, 1976, p. 14), changes in landscapes occur through the occupation and use of land. The geographer Junqueira adds that:

> Around 10,000 years ago, humans began to apply agricultural and livestock techniques that allowed them to attach themselves to the land, adding value to it that was unknown until then. Extensive areas of the earth's surface began to undergo profound changes, as happened in Europe with the replacement of forests with agricultural or grazing areas. As a consequence of this agricultural aptitude, thanks to agricultural surpluses, labour was released from the countryside for the incipient polis, city states, in Mesopotamia, between the Tigris and Euphrates (present-day Iraq), India, China, Greece and Egypt. From then on, human beings began to develop their geographical space in a more elaborate way (2003, p. 49).

Small human settlements were emerging: as examples, the hydraulic civilisations of the fertile valleys, periodically flooded by the river basins of Egypt (Nile), Mesopotamia (Tigris and Euphrates), China (Huang Ho and Yang Tse Kiang), India (Ganges and Indus), America (Amazon, Mississipi, Grande, Orinoco and Prata) and the coastlines of the Continents, from where villages, arraialias, towns and later cities were born, in which commerce and markets, industry and the advance of territorial, cartographic and naval knowledge appeared (ANDRADE, 1992, p. 20-22; AQUINO et al. 20-22; AQUINO et al, 1990, p. 21; FONTANA, 2000, p. 99-203; MAGALHÃES FILHO, 1983, p. 1-54).

According to Moreira (2007, p. 14), "in ancient times, geography was a cartographic record of peoples and territories. States, travellers and traders required information of a strategic nature from the geographer to guide them on their journeys through the spatial ways of life of each people".

Professor Andrade says:

> Eastern peoples not only developed empirical knowledge of geography, but also made observations and established mathematical studies that gave rise to systematic knowledge of the world. We can say that geographical ideas, in coexistence with those of other sciences, developed from the practical knowledge of exploring the Earth and the observations of travellers, alongside the systematisation of thinkers, philosophers and mathematicians (1992, p. 22).

Thus, teachers Silva and Fioreze (1999, p. 252-253) show that:

> The growth and expansion of humanity on the earth's surface, certainly guided in its

> march by the approximation and convergence of continental units, favoured the development of empirical and intuitive geographical activity, which emerged with the guides in prehistoric times and evolved from the stage of naming places and spatial phenomena to the stage of inventories.
>
> Knowledge was organised around solutions to everyday problems and the relationship between those who knew and those who learned was the construction of primitive knowledge and probably the genesis of the teacher-student relationship.

From then on, man began to organise his geographical space in a more elaborate and gradual way. Based on empirical knowledge, ancient civilisations have left us a wealth of geographical knowledge, mainly of a cartographic, territorial and naval nature.

Early geographical knowledge accompanied the expansion of ancient peoples into new territories. These peoples of the Mediterranean islands, especially the Greeks, developed and improved trade. They extended their borders, conquered, learnt about and recorded the knowledge of other lands for future use. Thus, today we have the reference documents that are the first maps (périplos and portulanos) with descriptions of voyages and maritime expeditions (ANDRADE, 1992, p. 2324; SODRÉ, 1976, p. 14-21).

Silva and Fioreze (1999, p. 253) state:

> In ancient times, the recognised geographical contribution of the Greeks, which also resulted from the accumulation of the legacy of the Mesopotamians, Phoenicians, Egyptians and Chinese, especially their nautical, cartographic and descriptive knowledge of the features of the areas where they carried out their commercial activities and practised agriculture, made them the people who exerted the most notable and decisive cultural influence on Western civilisation. What is clear to posterity is the concern they had with the localisation and description of places and the conception and practice of determinism.

For Sodré (1987, p. 14), "from this angle, it is generally accepted that the Greeks were the first to systematically record geographical knowledge" and "it was the Greeks, in fact, who baptised knowledge about the Earth's surface as Geography". Moraes (1998, p. 32) confirms that "the label Geography is quite old, its origins going back to Classical Antiquity, specifically to Greek thought".

Geographical genesis dates back to Ancient Greece, and has its initial parameters with scholars and thinkers of Natural History and Natural Philosophy. A large part of the so-called Western World was known and dominated by the Greeks, in particular the eastern Mediterranean, involving the Adriatic, Ionian, Aegean and Black Seas.

According to Broek (1972, p. 20),

> Any scientific endeavour depends first and foremost on careful observation. The Greeks were masters of this. They described the layout of the Earth and the character and customs of its inhabitants. The Greeks not only made descriptions of places (the so-called topographies) but also tried to explain them.

The author goes on to say that:

> We owe our knowledge of Greek geographical thought largely to two masterful compilers of the Roman era (Strabo and Ptolemy), who in their different interests clearly represent the two main aspects of Classical Geography (1972, p. 20).

One of them was Strabo, the father of Geography. In his speculation on the shape of the Earth, he wrote a work called GEOGRAPHY, in which he discussed his experiences in the lands of Galicia, Brittany, India and from the Black Sea to Ethiopia. The other was Ptolemy, who wrote GEOGRAPHIKE SYNTAXIS (in Arabic called ALMAGESTO), which emphasised his studies more towards mathematics, accompanied by projections, with tables of latitude, longitude and calculations of the day according to the distance from the equator (SODRÉ, 1976, p. 16-18; BROEK, 1972, p. 21, PEREIRA, 1989, p. 48; MOREIRA, 1994, p. 17; MORAES, 1998, p. 33).

In the 4th century BC, the Greeks observed the planet as a whole and the entire celestial vault. Through philosophical studies and astronomical observations, Aristotle was the first to take credit for conceptualising the Earth as a sphere with very precise measurements. Herodotus, the father of History, also known as one of the progenitors of Geography, carried out historical studies with an emphasis on the geographical context (JUNQUEIRA, 2003, p. 50; SODRÉ, 1976, p. 15-16, RODRIGUES, 2008, p. 43-46).

The Greeks left valuable information in their writings that told of their geographical experience, such as the detailed studies carried out on the River Nile during its period of natural overflow (RODRIGUES, 2008, p. 40).

For Sodré,

> The Romans extended this geographical heritage, and the conquest operations they carried out gave rise, in some cases, to descriptions of the areas they had then ravaged and subjected. The Romans were conquerors, but they still left it to the subjected Greeks to accumulate geographical knowledge (1976, p. 17).

In Andrade (1992, p. 27),

> The Romans, however, being very pragmatic, sought to develop the organisation of their Empire as much as possible, as well as trade in the dozens of provinces that made it up. Hence the greater importance they gave to descriptive geography and the lesser concern for mathematical geography, which was left to the wise Greeks.

The fall of the Roman Empire led to the "passing of the baton" of geographical knowledge from the Greeks to the Arabs. From then on, many reports and works, including the work of Ptolemy, were translated from Greek into Arabic.

Andrade (1992, p. 28), points to a question about the Roman fall:

> It was the expansion of Christianity, which became the official religion of Rome in the fourth century, that greatly emphasised the power of the monks. In their monasteries, the monks devoted themselves to the most diverse studies and sought to find the whole truth in the Holy Bible. Their power would become greater after the fall of the Roman Empire in the West in the 5th century, when they became the holders of European culture in the face of peoples classified as barbarians and uncultured. This dominance of biblical principles would push back a series of "scientific truths" accepted by the Greeks, returning to old theories, such as those that denied the sphericity of the Earth.

In Sodré, there are other issues:

> The collapse of Roman slavery and the merging of the gentile communities of the so-called "barbarians" into colonisation took place during a long period of turbulence that was matched, in the realms of culture, by stagnation or a reduction in activities that resembled it. And Christian Europe, when European and Christian were synonyms, was then faced with the extraordinary Muslim expansion, of which the Arabs were fierce bearers. And these are the heirs of Greek culture, which they assimilate and pass on

(1976, p. 19).

Moreira (1994, p. 16-17) raises other approaches:

> With the rise of Rome and the subjugation of the known peoples and lands, including the Greeks themselves, we see the consolidation of the slave mode of production. Thus, the aspect of democratic struggles went into decline. With the Romans, geography was restricted. It became a weapon in the formation of the vast and expanding empire, the largest of ancient times. An eminently militaristic state, the Roman Empire subjected knowledge to expansionist ends. Therefore, as dominators over the Greeks, the Romans needed to nourish themselves with the knowledge they produced. Geography's fate would be the same as that reserved for knowledge as a whole. The Geography that will develop will be the one that we will see serving the State, conceived as reports and maps, and will thus pass into history as Geography.

Junqueira reports that:

> It has to be said that this geographical knowledge built up by the Greek, Roman and Egyptian philosophers, among others, was, throughout the Ancient Age, a dispersed knowledge, which did not favour the formation of a cathedral knowledge, capable of guiding a concrete discipline. However, its resonance still resonates in today's geography, despite having been overshadowed by the scrutiny of the Catholic Church in the Middle Ages (2003, p. 51).

We can see so far that fieldwork was linked to expansionism, to the conquest of new lands. This shows that fieldwork, sometimes planned, sometimes unplanned, fulfilled different purposes.

We note that with the entry of the Bible into the Middle Ages, "the dogmatic justification of the Christian faith replaced free intellectual enquiry" (BROEK, 1972, p. 22) and scientific knowledge suffered regressive impacts in Medieval Europe (RODRIGUES, 2008, p. 49).

According to Broek (1972, p. 22), "the image of the world was made according to the

Greek thought, when contrary to Christian doctrine, had to be suppressed as pagan". This contrasting picture of Greek ideas and biblical thought meant that "during the 8th century and the following centuries, a new interest in knowledge spread throughout the Islamic world" (BROEK, 1972, p. 22).

According to Moreira (2007, p. 14),

> In the Middle Ages, the influence of the Church led Geography to be a form of vision that referenced the biblical imagery of a world created by God in his image and likeness. Medieval geography was therefore an extension of the Bible and the geographer was a cartographer of the fantastic.

Thus, "in Muslim universities, from Persia (now Iran) to Spain,

teachers studied the Greek legacy" (BROEK, 1972, p. 22). For Andrade,

> Some Arabs travelled far and studied not only the natural conditions and resources to be exploited in the areas they travelled, but also the institutions and customs of the peoples they dominated. These writers left notable works, which are still of great interest to geographical scholars today. Naturally, they were not specifically concerned with geography as such, but studied the characteristics of nature and the ways in which man could better exploit it and exploit other men. Among the most widespread Arab writers today are AL EDRISI, IBN BATUTA and IBN KHALDUM. All three were travellers, good observers, traders, believers in Allah and the Prophet, but they were men who reflected the society of their time and described nature and the problems linked to its

> exploitation (1992, p. 30-31).

In the Middle Ages, also known as the Dark Ages, with the Arabs, Geography played a major role due to the territorial conquests of the Byzantine Empire in Asia Minor and North Africa, the lands of Syria, Palestine, Mesopotamia (now Iraq), Iran, India and the Iberian Peninsula (Portugal and Spain).

According to Silva and Fioreze (1999, p. 253), "medieval Arab civilisation, essentially mercantile, found in commercial reasons and in the new religious belief, Islam, the impetus for extending its geographical horizon eastwards to India and westwards to the Iberian Peninsula".

In the 14th and 15th centuries, the Arab Empire suffered territorial dismemberment and bitter commercial control due to conflicts between the Turks and the Mongols. These were the same Turks who conquered Constantinople (now Istanbul) and destroyed the Byzantine Empire. The Arabs tried to extend their territory beyond the Pyrenees and were defeated by the French army. According to Andrade,

> As for the Arabs, after the preaching of Mohammed and the conversion of the peoples of the Arabian peninsula to Islam, they began, based on their new religious belief, to wage wars of conquest and conquer the lands of the Byzantine Empire in Asia Minor and North Africa. In just over a century, they dominated the old Orthodox Christian civilisations of Syria and Palestine and conquered North Africa, all the way to the Atlantic coast. After conquering Morocco, they crossed the straits linking the Mediterranean to the Atlantic and dominated the entire Iberian Peninsula. They tried to expand beyond the Pyrenees, but were defeated by the French at the Battle of Poitiers. To the east they dominated Mesopotamia and brought Islam to Iran and India (1992, p. 30).

With the end of the Middle Ages, according to Silva and Fioreze (1999, p. 253-254),

> There are no followers of this historian's thought (Ibn Khaldum) because it would correspond to the end of the Byzantine Empire, the Turkish and Mongol invasions, the
>
> the expulsion of the Arabs from Europe and the decadence of Arabic science. However, this does not detract from the fact that he was the bridge between the knowledge of the East and the West, which was taken over by the philosophers and scientists of Christianity.

The same authors go on to say:

> At the end of the Middle Ages, many factors contributed to the great changes that generated modern times and whose influence was felt in the development of a new economic, political, social and cultural order, leading to the evolution of geographical thinking (1999, p. 254).

As Rodrigues (2008, p. 55) puts it:

> It can be said that geographical knowledge in the Middle Ages suffered discontinuity in relation to the Old Ages due to the period of wars, invasions, the fall of the Western Roman Empire, the immobility of the population, the disruption of trade and travel, the emergence of the feudal mode of production and scientific knowledge under the influence and domination of the Church. But at the end of the Middle Ages, a new phase began for scientific knowledge, specifically geographical knowledge, with the changes of Modern Times.

Cidade (2001, p. 106) records that from the 5th to the 15th centuries,

> Medieval Europe, made up of different territories with varying degrees of cohesion around national projects, generally had feudal characteristics and was largely under the hegemony of the Catholic Church. In this amalgam, the distinction between worldviews and visions of nature according to different societies is unclear, since the doctrine of the Church and the Hebrew-Christian tradition predominated. In Europe, by imposing its values, religion even forced thinkers to make real setbacks, unlike the Arab world, which was able to value the baggage accumulated by other civilisations, such as Greece.

The "Renaissance brought, as in other sectors, a return to classical geographical thinking" and the 15th century saw "the rise of Portuguese and Spanish explorations, culminating in the voyages to India and America" (BROEK, 1972, p. 23).

Professor Cidade (2001, p. 107) states that "after the long period of hibernation of autonomous knowledge, Europe in the 16th and 17th centuries became the seat of a flowering of culture and the arts, with the revival of aesthetic values from the classical era, the Renaissance".

According to Moreira (2007, p. 14),

> In the Renaissance, geography was a form of cosmology designed to help conceive of the world as a great mathematical-mechanical system. And the geographer is transformed into a cartographer of the movement of celestial bodies in their geodesic rebounds on the earth's surface, endorsing a natural and desacralised view of the world.

Silva and Fioreze are quite categorical about the Renaissance:

> Philosophically, the Renaissance inaugurated a new logic, based on moving to the limit in order to think about the infinite, leaving the finite, closed world behind, in opposition to the scholasticism of the Middle Ages, and trying to recover Greek ideas through the
>
> a return to classical humanist values, the great hallmark of the Renaissance. It was from the Renaissance that the foundations of the scientific method emerged, proposed by DESCARTES (1596-1630) and centred on experience and rational intuition, opposing the scholastic thinking that had prevailed until then.
>
> In modern times, Europe underwent the most extraordinary expansion in history as a result of its mercantile movement and which inaugurated the capitalist mode of production. As a contribution to geographical science, in addition to spatial expansion, doubts about its configuration and surface were elucidated (1999, p. 254).

In their eagerness to know, discover, observe and experiment, the men of the period stretching from the 15th to the 17th century produced "ingenuities" for cartography and set out on the seas, tracing routes, mapping, based on the teachings promoted by the naval schools of Sagres (Portugal) in 1415 and Seville (Spain) in 1503, searching for the lands of America, Asia, Africa and the Australian archipelagos (RODRIGUES, 2008, p. 59; SILVA and FIOREZE, 1999, p. 59). 59; SILVA and FIOREZE, 1999, p. 254).

Various travellers (Vasco da Gama in 1498, Ferdinand Magellan in 1519, Christopher Columbus in 1492, Bartholomew Dias in 1488, Pedro Álvares Cabral in 1500, James Cook in 1770 and many others) broadened the known geographical horizon. Geographical description, mapping and cartography, driven by the need to expand, made great progress. It was the time of the Great Navigations, which explored and exploited the raw materials of the new lands (COELHO and TERRA, 2001, p. 12-13, RODRIGUES, 2008, p. 58).

Andrade (1992, p. 37), defines that: "hence the stimulus to expand the known horizon and the organisation of expeditions that sought new lands, new peoples and new merchandise with which to trade in distant lands".

Cidade (2001, p. 107) emphasises that:

> In England, capitalism was emerging, beginning the transition from an agrarian society to what would become urban and industrial life. Some of the main countries in Europe at the time, such as Spain, Portugal, France and England, had inaugurated the phase of discoveries and laid the foundations for colonialism, as a possibility for the expansion of commercial or mercantilist capitalism.

The views of Silva and Fioreze (1999, p. 254) show that:

> With a view to commercial and political activities in the newly annexed areas, the development of scientific knowledge was provided, not only with regard to navigation itself, but also with regard to astronomical, meteorological, climatic and tropical flora and fauna knowledge, which aroused great scientific interest and gave rise to the precursor studies of Geography as a science.

The new developments that emerged between the 15th and 18th centuries contributed to the process of systematising geography, which only became a science in the 19th century, thanks to the ideas of researchers Humboldt, Ritter and Ratzel.

From the Renaissance onwards, in the 15th and 16th centuries, scholars began to research man himself as a rational being who was superior to other beings. The concept of Humanism came into play.

By the 17th century, geographical knowledge, sparse and interconnected with the various related sciences, was already quite numerous and had acquired a certain depth. Examples are varied and in various fields of knowledge stemming from the contributions of the Greco-Arabs, Copernicus, Galilei, Kepler, Steno, Agricola, Da Vinci, Avicenna, Leibniz, Montesquieu, Newton, Darwin and others (ANDRADE, 1992, p. 44-45; LEINZ; AMARAL, 1989, p. 3-9; RODRIGUES, 2008, p. 60-63).

In the case of Geography, the notable figure was Bernardo Varenius (lived in the first half of the 17th century), who left a fundamental work in 1650, entitled GENERAL GEOGRAPHY (ANDRADE, 1992, p. 45; MORAES, 1998, p. 33; SILVA; FIOREZE, 1999, p. 254; BROEK, 1972, p. 24; CIDADE, 2001, p. 109).

The work of Bernardo Varenius has merits that Andrade (1992, p. 45) highlights:

> Varenius' great value comes from the fact that he united general, mathematical geography with descriptive, humanist and literary geography into a single whole and that he described and interpreted the forms and phenomena described, indicating cause and effect relationships. Varenius' work needs to be better publicised and discussed by those who work in geography today.

In the 17th century, the Enlightenment emerged with proposals in defence of social, economic and political freedoms, science and rationality, and criticism of the studies accepted by the Church.

Professor Moreira (2007, p. 14) states that:

> Between the Renaissance and the Enlightenment, geography doubled: on the one hand it was once again a cartography of the fantastic, but this time to highlight the imagery of a rational Europe in contrast to a world of barbarians that European reason had to conquer and civilise, and on the other hand it was a cartography of precision, aimed at the practical

> goal of guiding the naturalists and navigators who set out to conquer the unknown world. The geographer is thus a mixture of traveller and naturalist, whose role is to organise the exotic world from the outside according to European reason.

The same author adds:

> The Enlightenment 18th century, marked by the Industrial Revolution and the rise of the bourgeoisie to the status of ruling class, is the consolidation of this geography and this profile of geographer. That's because it was the century that called for a geography and a geographer who mapped the world with the mathematical rigour of locating and cubing the resources that the new economy decreed to be the top priority. Geography then becomes the science of large spaces and the geographer a specialist in the theory and practice of localisation (2007, p. 15).

According to Andrade (1992, p. 37),

> The thirst for wealth and the intensification of exchange between the West and the East led to cultural development and the spread of instruments that would be of great importance in the economic and social transformations that would take place in the 15th, 16th and 17th centuries, in the so-called Modern Times, and which would shake up political and social structures in the 18th and 19th centuries.

From this perspective, the cartographic base and the instrumental apparatus associated with the fieldwork resource worked strategically for the knowledge of new localities. Therefore, corroborating the ideas of Moraes (1998, p. 34-38), the systematisation of Geography took place at the beginning of the 19th century, according to four assumptions: "1) the effective knowledge of the real extent of the planet; 2) the existence of a repository of information on various places on Earth; 3) the improvement of cartographic techniques; and 4- philosophical and scientific changes"

The first refers to the work carried out during the period of the Great Navigations; the second is a database of physiographic and social materials from the new lands; the third concerns cartographic materials (charts, maps, routes, coordinates, atlases, etc.) and the last proposes a more rational explanation of the world, breaking with the religious view of the time (MORAES, 1998, p. 34-33; SOUZA, 2008, p. 30-31; JUNQUEIRA, 2003, p. 51-55). 34-38; RODRIGUES, 2008, p. 58-59; SOUZA, 2006, p. 30-31; JUNQUEIRA, 2003, p. 51-52).

Researcher Souza emphasises that:

> At the beginning of the 19th century, these historical presuppositions for the systematisation of Geography had already been outlined. Most of the places on the planet were already known. Europe was articulating a space of globalised economic relations, the development of trade brought the most distant places into contact and, therefore, the Geographies of Places became a necessity for the economic control of the world at that time (2006, p. 32).

It should be noted that Geography, before its scientific systematisation, was given various labels: "Non-academic Geography", "Unofficial Geography" and "Prehistory of Geography". In Moraes, there is a characterisation for the period known as unofficial geography:

> Thus, until the end of the 18th century, it is not possible to speak of geographical knowledge as something standardised, with a minimum of thematic unity and continuity in its formulations. The following are referred to as geography: travel reports, written in a literary tone; compendiums of curiosities about exotic places; arid statistical reports from administrative bodies; synthetic works, grouping together existing knowledge about natural phenomena; systematic catalogues about the continents and countries of the

> globe, etc. In fact, this is a whole period of dispersal of geographical knowledge, where it is impossible to speak of this discipline as a systematised and particularised whole (1998, p. 33-34).

Geographical knowledge passed through the hands of the Greco-Arabs, who managed to give "ingenious" impulses when they made a strong contribution to a scientific development path among the Germans and French at the end of the 19th century.

Souza (2006, p. 31) states that "all these material conditions for the systematisation of Geography were forged in the process of advancing and dominating capitalist relations".

Cidade (2001, p. 110), tries to show that:

> While parts of Europe, such as Germany, continued to live under feudal rule, capitalism was advancing, already in the phase known as competitive, characterised by the development of industry, the formation of an internal market and international trade, with England at the forefront. The expansion of capitalism into new territories, symbolised at the end of the 18th century by the end of feudal rule in France, was one of the most important processes in Europe at the time. The intellectual foundations of this change were rooted in the Enlightenment, a cultural movement in the phase between the English Revolution (1688) and the French Revolution (1789).

Reaffirming this, by the beginning of the 19th century, the geographical assumptions were already sufficiently structured, as a Contemporary Geography began to emerge under the "influences of Positivism advocated by both Comte and Kant" (SODRÉ, 1976, p. 27; SILVA; FIOREZE, 1999, p. 255; ANDRADE, 1992, p. 49). With this in mind, Cidade (2001, p. 115) outlines that "in the 19th century, Geography established its prelude to modernity, inaugurated by its general systematisation".

Junqueira (2003, p. 53-54), infers:

> The process of evolution of geographical thought, like that of several other sciences, occurred in parallel with the process of consolidation of Geography as a particular and autonomous science. It was an unfolding of the transformations brought about in social life by the emergence of the capitalist mode of production, which saw it as an instrument for achieving this process of building the capitalist mode of production in certain European countries, especially Germany and France.

For the same author,

> Germany and France are emblematic of the consolidation of capitalism in Europe. If there is one country that can be considered the homeland of contemporary geography, it is the Germany of Humboldt and Ritter. It was in Germany that the first institutes, the first professorships, the first methodological proposals and, consequently, the first currents of Geographical Thought emerged (2003, p. 54).

The qualitative leap in geographical knowledge took place in Germany at the end of the 19th century, at a time when there were disparities in the development of the capitalist system in England, France and Germany. The capitalist question was very well resolved among the English, well advanced with the French and among the Germans, the process happened late, because the problem at the time was internal unification (PEREIRA, 1989, p. 91).

Moreira's exposition is quite providential and clarifies that:

> It was among the Germans that, around 1754, Geography began its journey towards scientific status. The steps in this direction are already clear in the discussions between

> the two paths that emerged at the time: "Political-Statistical Geography" and "Pure Geography". The former continues what geography had been doing since the time of Strabo. The second emphasises the question of the natural limits of a territory, which will emerge in the 19th century with Ratzel in particular. Both took on the major problem posed at the time for the development of German capitalism: overcoming the backwardness of the more advanced levels of capitalism in England and France, and solving the domestic problem caused by the acceleration of capitalist development, i.e. the necessary unification of German territory (1994, p. 20-21).

Moreira (1994, p. 22) goes on to emphasise that "if, for English and French capitalism, the role of geography was to make colonial expansion possible for them, for German capitalism its role would be to provide answers to still preliminary questions: German unity". With regard to the German question, I endorse what the geographer Rodrigues says, adding some very important details:

> Germany experienced three conditions at the beginning of the 19th century: firstly, a territory fragmented into dozens of small fiefdoms; secondly, the desire for unification; and thirdly, imperialist expansionist ideals, combined with industrial capitalism.
> Germany was then in a unique situation:
> a) the absence of a nation state;
> b) the absence of an absolute monarchy, a form of government typical of the transition period in Europe;
> c) power in the hands of the landowners (the junkers), in a feudal structure;
> d) the lack of a strong economic centre to organise the territory;
> e) border disputes, between the fiefdoms and with neighbouring non-Germanic countries;
> f) economic backwardness of numerous German "states".
> The unification of Germany was led by Prussia and the formation of the German national state needed stimulus, which meant that the geographical discourse took on a sense of homeland through territorial identity (2008, p. 73).

German unification took place in 1871, under the aegis of Prussian Prime Minister Otto Von Bismarck, "unification will be achieved by iron and blood", after several confrontations with Denmark (1864), Austria (1866) and France (1870). In the conflict with the French, the Germans emerged victorious and annexed the French territories of Alsace and Lorraine, rich in coal and iron ore (RODRIGUES, 2008, p. 74).

According to Moraes (1998, p. 47), "Geography emerged in Germany, where the question of space was paramount". The favourable political and economic environment provided the conditions for the scientific study of geography in Germany. And so the Germanic School of Geography emerged, also known as the Determinist School, because several of its supporters believed that the natural environment governed man's actions and relationships. The main exponents of this school of thought were the Prussians Alexander Von Humboldt (1769-1859), Karl Ritter (1779-1859) and Friedrich Ratzel (1844-1904). Professor

Moraes (1998, p. 52) pointed out that "while Humboldt and Ritter experienced the emergence of the ideal of German unification, Ratzel experienced the actual constitution of the German national state".

For the three illustrious Prussians of the German School of Geography, Professor Andrade (1992, p. 51-54) points out some of the characteristics:

> 1- Alexander Von Humboldt was a Prussian nobleman who dedicated himself to

studying the natural sciences, especially botany. Very curious, he became a great traveller and explored Europe, Central and North Asia and Latin America. Although he was a naturalist, he was very curious about man and his social and political organisation, believing that this was closely related to natural conditions.
2- Karl Ritter was not a traveller, an explorer, but a great reader and an excellent expositor. He endeavoured to explain the evolution of humanity by linking it to the relationship between people and the natural environment, above all by describing society. He recognised that the whole was formed by the sum of its parts and that the sum of local particles could be used to formulate general laws valid for the entire surface of the Earth.
3- Friedrich Ratzel became famous for emphasising man in his geographical formulation. Living in Germany and having witnessed its unification under the aegis of Prussia, he formulated a geographical concept that corresponded to the expansionist desires of the New Empire. Devoting himself to the natural sciences, especially anthropology, he viewed man as an animal species and not as a social element, trying to explain the evolution of humanity within Darwin's postulates.

With regard to the naturalist Humboldt, Capel (1981, p. 5) makes the following comments:

> Almost all treatises on the history of geography agree that Alejandro de Humboldt is the father of modern geographical science. His work was undoubtedly decisive in shaping many geographical ideas, particularly in the field of physical geography.

Then the teacher says:

> In the case of Carl Ritter, this is idle, since he was undoubtedly a geographer who became Professor of Geography at the University of Berlin. His work focuses directly and fundamentally on the study of the relationship between the earth's surface and human activity (CAPEL, 1981, p. 41).

Again, the geographer explains:

> The essential ideas of Ratzel's anthropogeography come fundamentally from the convergence between his organic or biogeographical conception, his concern for the problems of diffusion and migration, and the ethnographic tendency (CAPEL, 1981, p. 286).

However, Capel (1981, p. 79) conjectures that "in reality Humboldt and Ritter would be, in any case, 'precedents' but not 'founders' of Contemporary Geography".
Therefore, for Andrade (1977, p. 6), there is no doubt about the roles of Humboldt and Ritter, because "the founders of Geography did not invent it from scratch". In this way, Azevedo (1959, p. 54) says of Humboldt: "his words and writings were oracular to his contemporaries", contributing to the most diverse sectors of human knowledge.

With regard to the German geographer Ritter, Professor Muller (1959, p. 82) states: "Carl Ritter was, for a time, a pioneer of Modern Geographical Science and of university teaching in this subject, since he held the first chair of Geography at the University of Berlin". The researcher Pereira (1989, p. 37) argues that "it was first taught by Kant at the University of Konisberg from 1756 to 1796, but it was from Alexander Von Humboldt and Karl Ritter onwards that Geography became institutionalised within universities".

According to the historian Sodré (1976, p. 30),

> If there was a need to set a date for commemorative purposes, for example, marking the autonomy of Geography, its constitution as a specific area of knowledge and its systematised analysis, that year would be 1845, when Humboldt began publishing COSMOS.

Moreira (1994, p. 26) points out that "with Ritter, Geography-History took on academic and school form, that is, the conception of the world as an anthropocentrism, a unity whose starting point and purpose is man".

With the resolution of the "dispute" provoked by Capel (1981, p. 79), and the victory of the Germans over the French in the battle of 1870, there were internal changes in France on the question of geography.

Rodrigues makes some important points:

> It is essential to emphasise that France was defeated, humiliated, left with a huge war debt and also lost the territories of Alsace and Lorraine to Germany. The republican government therefore sought to recover, and one of the sectors it was very concerned about was education. Hence the need to reorganise teaching, placing greater emphasis on the subjects of Geography and History at secondary level.
> The long-standing rivalries between France and Germany increased with the loss of Alsace and Lorraine during the Franco-Prussian War. This fuelled the development of geography in France, as France's loss of the war was attributed not to the German army, but to geography, how it was taught and applied (2008, p.81).

On French soil, after 1870, criticising Ratzel's geography, the path was blazed for the creation of a French School of Geography, also known as the School of Possibilism, which believed in the possibility of reciprocal influences between man and the natural environment, based on the idea that man adapts to the physical environment, as well as modifying it. Paul Vidal de la Blache (1845-1918) was the illustrious geographer of this school.

For Moraes (1998, p. 68), "Paul Vidal de la Blache defined the object of the Geography as the relationship between man and nature, from the perspective of the landscape. He placed man as an active being, who suffers the influence of the environment, but who acts on it, transforming it". Andrade mentions that La Blache was

> Deeply embedded in the dominant political thinking, he carried out a series of regional studies, analyses that we could call micro-geographical, in which he tried to demonstrate that the environment exerted an influence on man, but that man had the possibility of modifying and improving the environment, giving rise to possibilism (1977, p. 7).

Campos (1996, p. 3), reports that:

> In 1892, Paul Vidal de la Blache created the chair of "Colonial Geography", later called "Géographie D'Otre Mer" and then "Tropical Geography". The purpose of serving the colonising power was already organised. The French government also commissioned him to draw up a new regional division of the country (the existing one dated from 1790), because the socio-economic reality, including the transport revolution, was different; he proposed a division into fifteen regions, each with its own capital.

The German and French schools of Geography emerged under the scrutiny, objectives and purposes of legitimising the practices of the states they served and are grouped together and nicknamed Traditional Geography, as they have descriptive characteristics and links to the positivism of Comte and Kant. According to Christofoletti (1985, p. 12), "the contributions and ideas presented by German and French geographers had a major influence on the development of this science in the first half of the 20th century".

Again on German soil, at the beginning of the 20th century, with the professor and geographer

Alfred Hettner (1859-1941), from the University of Heidelberg, carried out a study on the differentiation of areas, with the aim of explaining the differences in the portions of the earth's crust (JUNQUEIRA, 2003, p. 63; SILVA; FOIREZE, 1999, p. 257, RODRIGUES, 2008, p. 93, CHRISTOFOLETTI, 1985, p. 12).

Hettner's ideas were not widely disseminated in his time, due to the clash between possibilists and determinists. Only in the United States, in 1940, was Hettner's legacy taken up and emphasised by Richard Hartshorne, a geographer and professor at the University of Winconsin (MORAES, 1998, p. 84-92).

Hartshorne's thinking shows a Geography linked to a study of variations in areas, where the basic concepts of "area" and "integration" are formulated and an area would have several integrated processes, being an inexhaustible cradle of inter-relationships (MORAES, 1998, p. 88, ANDRADE, 1992, p. 78-79).

Hartshorne, in his studies, presents two forms of geographical approach: the Idiographic and Nomothetic. The idiographic study, "would be a singular analysis (of a single place) and unitary (trying to apprehend several elements), which would lead to a very deep knowledge of a particular place" and in the "nomothetic study, the researcher would stop at the first integration, and reproduce it (taking the same phenomena and making the same inter-relationships) in other places" (MORAES, 1998, p. 89).

It is worth noting that both Hettner and Hartshorne are illustrious representatives of the Rationalist School, whose "aim was to develop a less empiricist Geography and to favour deductive reasoning" (RODRIGUES, 2008, p. 93).

The thoughts of Moraes (1998, p. 91-92) point to the end of the cycle of Traditional Geography, stating that "the proposals of Hartshorne, on the one hand, and André Cholley and Maurice Le Lannou on the other, bring to a close the last attempts of Traditional Geography".

In the 20th century, new approaches were born and made it possible to see Geographical Science in broader terms. The object of study and the objectives of Geographical Science were given different definitions, depending on the moment and the ideological vision of scholars and thinkers and their links to socialist and capitalist economic systems.

In the aftermath of World War II (1939-1945), a period known as the Cold War, Atomic Blackmail or the Space Race, in the clash between the Americans and the Soviets, there was a desire to make Geography a more accurate and scientific study, more tangible as a discipline, which led to the introduction of Maths and Statistics as teaching resources. Thus, in addition to the exact sciences, two new techniques of substantial importance for Geography were beginning to be developed: the computer and satellites, giving focus to the applied discipline.

The labels are rich: the School of Quantitativism, the Pragmatic School and the Systemic/Modelling School, which embarked on technological and mathematical-statistical exaggerations, marginalising the genesis of information. These schools include researchers such as M. Philipponeau, Giuseppe Dematteis, J. P. Cole, Bryan Berry, Sotchava, Manley, Ian Burton,

Fred Schaefer, David Harvey, Chorley, Haggett, among others (CHRISTOFOLETTI, 1985, p. 16-32; MORAES, 1998, p. 100-111; JUNQUEIRA, 2003, p. 67-69; CAPEL, 1981, p. 378-381).

According to Capel (1981, p. 367), "from the 1950s onwards, Geography experienced a profound change in the English-speaking world, which gave rise to the so-called quantitative revolution, from which a new geography, a New Geography, emerged".

The Spanish geographer goes on to explain that

> "It is now emphasised that, since mathematics is the language of science, it must also be the language of geography. Therein lies, and not simply in the use of statistics, one of the essential motives leading to the mathematisation of geography, to the emergence of Quantitative Geography (1981, p. 386).

From "Old Geography" to "New Geography", Moreira (1994, p. 44-46) conjectures:

> So instead of describing the landscape, mathematisation takes its place. Instead of describing the morphology of the landscape, a rigorous typology of spatial patterns takes its place. Instead of field research, computers take over. Instead of subjective description, the descriptive objectivity of mathematical language takes over. Geography has therefore made a huge leap: from Positivism to Neopositivism. From one place to the same.

In agreement with Professor Christofoletti,

> The term "New Geography" was first proposed by MANLEY (1966), considering the set of ideas and approaches that began to spread and develop during the 1950s. The emergence of new approaches was part of the profound transformation brought about by the Second World War in the scientific, technological, social and economic sectors. This transformation, covering both the philosophical and methodological aspects, was called the "Quantitative and Theoretical Revolution in Geography" by IAN BURTON (1963). Although historical evidence can be found as far back as the 1940s, FRED SCHAEFER's contribution, in 1953, on EXCEPTIONALISM IN GEOGRAPHY: A METHODOLOGICAL EXAMINATION, chronologically marks the realisation of these renewing tendencies (1985, p. 16).

According to the same researcher (1985, p. 16-20),

> In an attempt to overcome the dichotomies and methodological procedures of Regional Geography, the New Geography developed in an attempt to encourage and seek a greater framing of Geography in the global scientific context. In order to give a general overview of the New Geography, we can specify some of its basic goals: a) Greater rigour in the application of scientific methodology - based on the philosophy of logical positivism, scientific methodology represents the set of procedures applicable to carrying out scientific research. What differentiates each science from the others is its object. Each science contributes to understanding the existing order and structure, and Geography's sector is spatial organisations. And from this perspective, two geographic works have gained greater prominence: EXPLANATION IN GEOGRAPHY, by David Harvey (1969) and AN INTRODUCTION TO SCIENTIFIC REASONING IN GEOGRAPHY, by D. Amedeo and R. Golledge (1975); b) Development of theories - the lack of theories explicitly set out in Traditional Geography has been vehemently criticised by numerous geographers. For this reason, under the paradigm of scientific methodology, the New Geography also sought to stimulate the development of theories related to the characteristics of the spatial distribution and arrangement of phenomena (the theories of Walter Christaller, Von Thunen, Losch, Alfred Weber); c) The use of statistical and mathematical techniques - the use of mathematical and statistical techniques to analyse the data collected and the spatial distributions of phenomena was one of the first characteristics that stood out in the New Geography. And its charisma was so great that it was reflected in the adjective used by many works, the name "Quantitative Geography". The most notable work is the recent publication by the IBGE on CURRENT TRENDS IN URBAN-REGIONAL GEOGRAPHY: THEORISATION AND QUANTIFICATION, organised by Speridião Faissol (1978); d) The systemic approach - the systemic approach serves geographers as a conceptual tool that enables them to deal with complex sets, such as those of spatial organisation. The concern to focus on geographical issues from a systemic perspective was a characteristic that favoured and boosted the development of the New Geography. For example, the introduction of the concept of GEOSSYSTEM by Soviet geographers made it possible to recompose and revitalise the field of Physical Geography (Sotchava, 1977) and the contribution made by Christofoletti (1979), who wrote ANALYSIS OF SYSTEMS IN GEOGRAPHY, should be highlighted; e) The use of models - closely related to the verification of theories, quantification and the systemic approach, the use and construction of models developed. For the geographer, the model is a working tool that should be used to analyse the systems of spatial organisations. As with quantification, the

construction and use of models should not be restricted to their simple purpose. But it is a means of better understanding reality. R. J. Chorley and Peter Haggett's work on MODELS IN GEOGRAPHY, published in 1967, is the classic contribution on the subject.

With regard to alternative geographical trends, Christofoletti (1985, p. 2032) emphasises:

> Basing its conceptual concerns on the theses of logical positivism, scientific methodology formalised itself around a few key positions, among which the following should be highlighted: a) Fruitful scientific knowledge is based on facts, on events collected in the empirical world; b) In order to be certain of knowledge, hypotheses must be verified, using the most diverse techniques of testing, and laws must be formulated. This type of certainty is provided by the experimental sciences. In more recent times, the criterion of refutability proposed by Karl Popper has been taken as the basic point for scientific methodology; c) The scientific procedure must always stick to contact with the experience of the empirical world, in order to avoid verbalism and error.
> However, it is normal and expected that reactions against the New Geography would emerge, seeking to follow other philosophical paths, which challenge and seek to replace the precepts of scientific methodology of a positivist lineage. Humanistic Geography (1960/1970) by Yi-Fu Tuan, Davis Lowentahal, Anne Buttimer and Edward Relph, Idealist Geography (1960/1970) by R. G. Collingwood and Leonard Guelke, Radical Geography (1960/1970) by John Hopkins, Simon Fraser, David Harvey, David M. Smith, Yves Lacoste and Richard Peet and Temporal-Spatial Geography (from 1970) by Torsten Hagerstrand and Alan R. Pred.

Both Rodrigues (2008, p. 116-120) and Andrade (1992, p. 118-121) show another geographical trend in the 1970s:

> In Ecological Geography there is no ideological identity between the various geographers on the solutions to be adopted in relation to the destructive impacts on the environment, but in common they advocate the preservation of nature. It's a movement in defence of the environment. Geographers who follow this orientation generally belong to other currents of geography, for example, Critical Geography, Humanistic Geography, Geography of Perception, Cultural Geography. In this geographical trend, we have the illustrious René Dumont, Ignacy Sachs, Jean Tricart, Manuel Correia de Andrade, Walter Leser, Warwick E. Kerr, Pakoff, Hilgard O'Railly Sternberg, Aziz Ab'Sáber, Carlos Augusto Figueiredo Monteiro, among others.

Thus, Radical, Critical or Libertarian Geography, from the 1970s onwards, whose exponents included P. Kropotkin, E. Reclus, Richard Peet, Yves Lacoste, Jean Tricart, Caio Prado Júnior, Orlando Valverde, Manoel Correia de Andrade, Milton Santos, Ruy Moreira, J. Dresch, M. Rochefort Bernard Kayser, Nelson Werneck Sodré, Pierre George and others sought to give a new perspective to geographical science, dealing with social, economic and political issues through Marxist approaches.

For Capel (1981, p. 403),

> Towards the end of the 1960s, this crisis translated into the proliferation of critical or radical movements, which developed in all the social sciences. It is in relation to all this that new currents of thought appear in the field of Geography, which lead the discipline down unprecedented paths, while at the same time making it possible to recover an important part of the historicist heritage.

In Garcia (1978 apud Christofoletti, 1985, p. 28-29), four guiding trends in Anglo-Saxon Radical Geography are outlined:

> 1- Anarchist line, centred on the University of Simon Fraser and the University of Clark, the latter emphasising the work of Richard Peet. This line traces its origins back to the

> pioneering work of Kropotkin and Reclus; 2- The popular-radical line, which is characterised by the geographers' direct contact with the populations of the areas and neighbourhoods to be investigated. The geographer participates and guides the population to solve their problems and outline their demands. The work of William Bunge (1971) is an example of this type of procedure; 3- The line with a Third World orientation, exemplified by the work of J. M. Blaut, aimed at proposing analyses of development and imperialism, among various other themes; 4- The Marxist orientation, based on the study of the works of Marx and Engels, the search for theoretical foundations and their application to socio-economic problems of spatial expression. The work of David Harvey is a prime example of this orientation.

According to Capel (1981, p. 434), "the discovery of this broad theme required new theoretical frameworks for analysis. This was the moment when Marxism revealed itself as an appropriate support for an alternative approach".

Christofoletti (1985, p. 31-32) shows that "it is up to the geographer to get to know the various trends, evaluate their positive and negative points, their advantages and disadvantages and consciously opt for one of them". It should also be emphasised that shouting for a geographical current does not mean that the chosen current is omnipotent, omnipresent and omniscient.

According to Moraes and Costa (1996, p. 120) "as we can see, the range of 'alien influences' on Geography is not small". From then on, a starting point is assumed,

> A clear choice of method provides elements for drawing up the work plan and its implementation. It provides a criterion of relevance in the selection of topics to be covered. It enables us to rethink Geography in the light of a solid parameter, based on an interpretation of reality that transcends and encompasses Geography itself (MORAES and COSTA, 1996, p. 116).

We can therefore say that the fieldwork had clear objectives: territorial reconnaissance, handling cartographic equipment, observing and describing the landscape and geopolitical use.

Next, we'll get to know a little about how fieldwork is viewed and worked with in the currents of Geography and in Brazilian Geography.

1. 2. Fieldwork in geographical currents

In this topic of the research, we are expressively interested in knowing clearly how fieldwork is discussed and used by each current of geographical thought. In this way, we will later be able to take a firm stance on the conceptions of teachers and students on the issue of fieldwork in Geography degrees in south-eastern Goiás.

It is therefore recognised by us geographers that geography became an autonomous science from the 19th century onwards, based on the work of Prussian geographers Humboldt and Ritter.

However, this is not to say that there was no knowledge or application of geography before the 19th century. Let's remember the work carried out by the Greeks, Arabs, Iberians and others.

According to Pedone,

> Fieldwork has become an essential element for a study to be recognised as geographical. At the same time, it seems that geography is recovering the tradition of explorations. Some authors agree that the explorations of Muslim scholars, Scandinavian travellers, Chinese travellers and the adventures of medieval Christians contributed to geographical knowledge. In addition, the European explorers of the 15th and 16th centuries helped to give greater coherence to knowledge about the surface of the globe. In the 18th century, we can mention the travels of Alexander Humboldt, or the overseas explorations financed by the Royal Geographical Society, among others; therefore, Geography has always been associated with travel and exploration (2000, not paginated).

During the "prehistoric phase" or "unofficial phase" of Geography, from Antiquity to the Middle Ages, geographical knowledge was linked to Cartography and Astronomy, as it was used to plot routes, pinpoint exploration resources, set up meteorological analysis points, among other aspects (ANDRADE, 1992, p. 12).

Oliveira emphasises that:

> Since ancient times, geographers have posed questions to nature that involve spatial problems. Initially, he was looking for ways to measure the planet. The first geographers were also geometers. Measuring and describing the Earth were seen together as the study concern of the same wise man, who sought to understand nature from a spatial point of view. The representation of terrestrial space arose as a result of this union of measuring and describing, carried out by groups of individuals who accumulated the functions of geographers, geometers and cartographers, and were concerned with the size and shape of the Earth.
> Earth, and above all by localising the events that took place on the earth's surface. Travel descriptions have always been perceptive accounts of the physical space visualised by the writer (1976, p. 55).

During this period, one of the great scholars was Strabo, who ventured to draw conclusions beyond the descriptions he made (ANDRADE, 1992, p. 12; SODRÉ, 1976, p. 16-18; OLIVEIRA, 1976, p. 55; BROEK, 1972, p. 21). Still Oliveira, "Herodotus, Strabo and others tried to describe what they saw or heard by linking the facts through cause-effect relationships, trying to represent the place where they occurred, to better clarify and situate the readers" (1976, p. 55-56).

According to Sodré,

> For a long time, Geography was thought to be an activity of observation, of purely collecting data. Hence its descriptive content. Later, with the accumulation of data gathered through observation, the intimate solidarity that unites things and beings, how they are composed, how they fit together, how they merge as a whole was characterised (1976, p. 7).

He also points out that:

> It was a time of great scientific journeys, with countless observations, data, materials and species being collected. From one of these journeys, Humboldt brought back the elements needed to organise Berghaus' Physical Atlas of 1836, in which the relationship between climate and vegetation was clearly shown (SODRÉ, 1976, p. 7).

He goes on to say

> The breadth of the field of observation and the extent of data collection ultimately led to some limitations in the field of Geography and, as in other fields, specialisations emerged. Despite this, Geography was left with the elementary method of which everyone has sad school memories. Previously, it was not understood that information is raw knowledge; it constitutes material for science, but it is not science (SODRÉ, 1976,

p. 7).

In the modern period, people began to look for deeper, more elaborate explanations of the relationship between physiographic elements and societies; thus, we have the precursor works of Bernardo Varenius, Kant and Montesquieu (ANDRADE, 1992, p. 12; BROEK, 1972, p. 24-25).

In the contemporary phase, according to Broek,

> While Kant's importance for Geography lies mainly in having provided the philosophical justification, Alexander Von Humboldt (1769-1859) and Carl Ritter (1779-1859), Germans too and equally erudite, modelled the substance of Geography, giving it a scientific form" (1972, p. 26).

Corroborating Oliveira (1976, p. 56-57),

> When the criterion adopted for classification was the chronological appearance of the science, situating Geography was a bit awkward. We all know that Geography has appeared as scientific knowledge since Greek Antiquity, as a direct product of Maths. Geography was born together with Geometry, the former being responsible for describing the Earth and the latter for measuring it. The Ionian, Athenian and Alexandrian geographers were also geometers. On the other hand, the so-called "Modern" Geography appeared in the 19th century. Faced with this impasse, where should Geography be placed historically: among the classical or modern sciences?

According to Wooldridge and East (1967, p. 15),

> It has been possible to investigate the early history of geographical thought in great detail. Nevertheless, there is a great gulf between modern geography and that of the classical and medieval periods. However, it is worth remembering that man, as a being capable of observing, comparing and reflecting, has always been a geographer, in a sense. In the remote past, his world and his environment were limited in space. This is still true for far too many people. However, it was necessary to cultivate a certain practical knowledge of one's surroundings, an elementary sense of the Earth, in order to live and survive anywhere on the planet.

In this brief overview, we can categorically state that fieldwork is one of the most traditional practical resources in Geography. It has always been used as a tool by man to get to know places, and this practical resource dates back to antiquity and contemporary geography. This tool is present in universities and primary and secondary schools, and is based on observation, description and explanation.

According to Oliveira (1976, p. 56),

> The work of geographers throughout history has always been concerned with the location of places, and above all with why phenomena occur here and not there. In other words, when faced with events, geographers try to describe how they are distributed in space; to explain spatial variations, i.e. the relationships between man and nature; and to organise space by dividing it into regions. Thus, localisation, relationships and regions have always been present in geographical studies.

For this approach, Santos (2008, p. 18) says that "description and explanation are inseparable". Carlos (2002, p. 164) points out, "A CITIZEN WHO DOES NOT THEORISE IS A SECOND-CLASS CITIZEN and the power of Geography is given by its ability to understand the reality in which we live, as Professor Milton Santos said at the time".

On the other hand, according to the same author, "Manuel Correia de Andrade wondered whether Geography should be mere intellectual reverie or whether it should provide the conditions for rationalising the organisation of Brazilian space, offering a contribution to the solution of Brazilian problems?" (CARLOS, 2002, p. 164).

Professor Andrade (1999, p. 55) states that "a Geography that distances itself from research, from field observations, that remains only in its theoretical aspects, is a useless Geography".

According to Cruz Neto (1998, p. 51), "fieldwork presents itself as an opportunity not only to get closer to what we want to know and study, but also to create knowledge based on the reality present in the field".

We can see that fieldwork that doesn't have well-defined geographical goals and doesn't contribute to the renewal of knowledge is like a well of salt water in the middle of the desert.

Completing the above reasoning,

> The extraordinary interest that geographic excursions, whether university or not, have aroused among us lately is not only the corollary of a healthy orientation in the teaching of geography. It is also the result of the efforts of those, starting with Delgado de Carvalho, who have seriously endeavoured to take the research into the science that Humboldt and Ritter created into the field (PEREIRA, 1943, p. 7).

Pereira emphasises that:

> The geographical excursion has its own objectives, in keeping with the way in which the purpose of geography is considered. One of them is to look at the totality of space, to look at all the different manifestations by studying the way in which they are particularly expressed in the landscape through the combination of locality and reciprocal interaction (1943, p. 8-9).

Thus, "an excursion offers us a multitude of opportunities to apply sound geographical teaching methods" (PEREIRA, 1943, p. 9), because in any place chosen for fieldwork "there is something to see, something to reflect on in Geography, because what we need is to know how to 'see', to know how to 'dialogue' with space" (SECRETARIA MUNICIPAL DE EDUCAÇÃO DE SÃO PAULO, 1992, p. 40).

We have in Wooldridge and East, that:

> Few geographers will fail to recognise, at least in general terms, the important role that fieldwork can and should play in the study of their subject. Beginners, however, may be filled with embarrassing doubts about the true nature and objectives of such work (1967, p. 171).

According to Andrade (1992, p. 13),

> The evolution of academic geography, with the contribution of numerous scientists who published their works in the late 19th and early 20th centuries, led Emmanuel de Martonne to define geography as "the science that studies the distribution of physical, biological and human phenomena over the surface of the Earth". This is a major evolution on the previous definition, since the study of this distribution is not only about describing, but also about explaining how and why the distribution is as it is. It is this explanation that gives Geography its scientific status.

Thus, some currents and schools of geographical thought should be highlighted in order to discuss the importance of fieldwork in each one of them, which is why the types of fieldwork need to be explored in greater depth.

So let's start with how the German school of geography deals with fieldwork.

1.2.1. The German School of Geography

Germany is considered to be the cradle of the institutionalisation of Geography (19th century), and home to the geographers Humboldt, Ritter and Ratzel, who contributed greatly to the science.

Alexander Von Humboldt (1769-1859), a scholar of the natural sciences and author of the works COSMOS (1845-1859) and FRAMEWORKS OF NATURE (1808), travelled and did fieldwork in Spanish America, the United States and Siberia. His geographical proposal consisted of observing, studying and describing the great physical and biological phenomena in their natural environment (WOOLDRIDGE; EAST, 1967, p. 20-21; ANDRADE, 1992, p. 51-54; BROEK, 1972, p. 26-27).

For Helferich,

> Today, you can feel Humboldt's spirit as far away as La Ciénega (south of Ecuador), and even beyond Latin America. From 1799 to 1804, Humboldt and his travelling companion Aimé Bonpland undertook what has been called "the scientific discovery of the New World", traversing a 9,500 kilometre stretch through what are now Venezuela, Colombia, Ecuador, Peru, Mexico and Cuba (2005, p. 16).

He also emphasises that, according to Humboldt, "nature never ceases to offer the most fascinating novelties for learning" (HELFERICH, 2005, p. 210). Humboldt's contribution to the sciences, especially geography, is immeasurable. For, according to Helferich (2005, p. 50-51), "the scientist, in other words, needs to become an explorer" and "this demanding methodology would in fact become known as HUMBOLDTIAN SCIENCE".

Agreeing with Sant'anna Neto,

> More than just providing information and discoveries about natural landscapes and their characterisation, Humboldt proposed a new way of producing knowledge. His conceptions of the natural landscape, of a living and dynamic nature and of its aesthetic, sensitive and unitary dimension, transformed and influenced contemporary thinking in the natural sciences and particularly in Geography. Practically all the natural science produced in Brazil from the Empire onwards (early 19th century) demonstrates the importance of Humboldt's work. Not only Physical Geography, but also Biogeography and Regional Climatology were founded on his postulates (2004, p. 43-44).

In this way, we can say that Humboldt's investigative spirit was crucial and fundamental to geography. The observation, description and apprehension of landscapes that Humboldt so appreciated became important methodological steps for the first geographers at universities in Germany and elsewhere.

Karl Ritter (1779-1859), philosopher, historian and geography lecturer, taught at the University of Berlin under Friedrich Ratzel, Élisée Reclus and Paul Vidal de la Blache. He was not a traveller, an explorer, but a great reader and speaker (PEREIRA, 1989, p. 117-120;

ANDRADE, 1992, p. 51-54). For Pereira, "due to his activities as a preceptor and later as a professor at the General War College and the University of Berlin, Ritter was linked to pedagogical concerns from an early age" (1989, p. 117-118). Continuing, "for him, Geography is essentially a historical discipline that has its own centre in the study of the relationships between the natural environment and the development of peoples" (PEREIRA, 1989, p. 118).

Pereira says:

> In ERDKUNDE, for example, he presents a detailed picture of the orographic and hydrographic structure of the African and Asian continents, in an effort to determine the possibility of life that the environment offers to the peoples who settle in it, the influence that this relationship has on historical events and, vice versa, the changes caused by man in the environment (1989, p. 118).

He also points out that "he opts for the historical approach and sees the earth's space as the theatre of history, considering that the greatest harmony between man and nature occurs at times of greater cultural development" (PEREIRA, 1989, p. 120). His work COMPARED GEOGRAPHY methodologically underpinned a geographical concern of a normative nature and defines the concept of a natural system as a delimited area endowed with individuality (MORAES, 1998, p. 48). For Sodré, "Ritter did not deny the empirical side of the method: geographical truth, in his view, lay in starting from observation to conclusion and not from opinion or hypothesis to observation" (1976, p. 35).

Finally, we note that Ritter's legacy in terms of geography is translated into a regional and anthropocentric perspective. Friedrich Ratzel (1844-1904) taught Geography at Leipzig University, travelled around Europe and America and, influenced by Darwin's theories, coined the term geographical determinism in his work ANTROPOGEOGRAPHY (1882) and said that the struggle between species was basically based on space, which would also apply to humanity (RODRIGUES, 2008, p. 75, MORAES, 1998, p. 55-60; ANDRADE, 1992, p. 54).

According to Rodrigues,

> In Ratzel's view, Geography is a science of synthesis, descriptive, empirical and works with observation and the collection of information in the field, seeking the relationship between phenomena, causal relationships from an inductive perspective, accepting the positivist conception. In other words, Geography uses the same methods applied by the natural sciences (2008, p. 77).

He also emphasises that "Ratzel defined geography as the study of the influence of natural conditions on humanity" (RODRIGUES, 2008, p. 76).

Drew emphasises the role of German determinism:

> The theory according to which natural conditions govern man's behaviour and even aspects of his character is called determinism or causality. It is a notion derived from the post-Drawin idea of man as a product of natural selection, by inexorable processes of nature. That nature obeys a grand plan, to which man must conform and thus prosper, is an entirely outdated thesis (1989, p. 4).

In Rodrigues' words, Ratzel "compares society to an organism that maintains strong relations with the soil in order to fulfil its need to survive" and from this he "develops the concept of living space, which consists of the balance between the population and the resources available for survival" (2008, p. 76).

According to Andrade,

> Ratzel's work, expressed in his two most famous books, ANTROPOGEOGRAPHY and POLITICAL GEOGRAPHY, despite the criticism it aroused, had a great influence on the development of Geography, emphasising the role of man and clearly demonstrating the political and social character of this science (1992, p. 56).

Ratzel's ideas would soon be used by the Nazi scientists led by Adolf Hitler to justify territorial expansion and the Holocaust, based on the geopolitics adopted by the European superpowers of the period (MORAES, 1998, p. 56).

Next, we'll look at how fieldwork is viewed in the French geographical school.

1.2.2. The French School of Geography

Paul Vidal de la Blache (1845-1918), founder of the French School of Geography, was largely responsible for removing the focus of geographical discussion from German soil, shifting the geographical dialogue to the French scene which, in turn, became opposed to the German school of Humboldt, Ritter and Ratzel (RODRIGUES, 2008, p. 79-82; MOREIRA, 1994, p. 34-39). For Moreira, "the 'French school' was born out of the climate produced by France's defeat by Prussian Germany in the war of 1870" (1994, p. 34).

According to Andrade (1992, p. 70), Paul Vidal de la Blache,

> Having been a disciple of Ritter, he was imbued with geographical concerns and accepted, to a certain extent, the influence of the environment on man. So much so that he never considered geography to be a social science, but rather a natural science, "of places".

Paul Vidal de la Blache, its greatest possibilist representative, wrote countless articles, a book UNIVERSAL GEOGRAPHY, founded and coordinated the magazine ANNALES DE GÉOGRAPHIE and one of its active principles was precisely the criticism of Ratzelian geographical determinism, which he accused of having introduced political discussions into geography (ANDRADE, 1992, p. 70; MORAES, 1998, p. 73-75).

For Drew (1989, p. 4), "there is an alternative thesis, which is that of possibilism; man is not passive, but a great geographical agent, capable of acting on the environment and modifying it, within the natural limits of space and development possibilities".

According to Moraes, "Vidal de la Blache defined the object of Geography as the relationship between man and nature, from the perspective of the landscape. He placed man as an active being, who suffers the influence of the environment, but who acts on it, transforming it" (1998, p. 68).

For Moreira,

> In this way, the "tree" of Geography branched out fantastically. Likewise, the difference in emphasis between Humboldt and Ritter regarding the scale of reference - for Humboldt the whole and for Ritter the region - will be translated by the French as a separation between General Geography (Humboldt) and Regional Geography (Ritter). The region

> would be consecrated as the object of Geography, to the study of which French Geography would turn almost entirely. And it gained such an expression that for a long time doing geography work meant producing a monographic work (1994, p. 38).

For Broek,

> As Vidal, in rejecting environmental determinism, often spoke of "environmental possibilities", his point of view was known as possibilism. But Vidal didn't mean that man is a free agent for whom everything is possible. He fully recognised that man's choice is severely limited by his society's value system, its organisation, technology, in short, by what Vidal called man's "genre de vie" (way of life or genre of living) (1972, p. 38).

Corroborating Andrade (1992, p. 70),

> Paul Vidal de la Blache's basic ideas fought against linear evolution. Preached by the positivists and evolutionists of the 19th century, it relegated globalising theoretical concerns to the background and focused on fieldwork, valuing intuition, the "clinical eye" of the geographer. He contrasted the general with the regional, leading his disciples to carry out field studies limited to small areas, regions, taking into account physical aspects and superimposing human and economic aspects on them.

Rodrigues reports that:

> Another important element in Paul Vidal de la Blache is the empirical orientation of geography, as he emphasises the importance of geographical work being carried out mainly through direct observation of reality, the need for geographical excursions as pedagogical work, in other words: the outdoor school should guide the geographical spirit. Reality existed independently of the observer, it had to be found in the field, emphasising the observation of the landscape in order to delimit and study the region (2008, p. 85).

Thus, we have yet another essence of fieldwork from a Lablachian perspective, where the focus is on the regional. These observations establish fieldwork as a tool for recognising one's own regional space for integrative actions between the physical, human and economic aspects that have spread to many locations, including Brazil.

Next, we'll look at how the British School of Geography deals with fieldwork.

1.2.3. The British School of Geography

Hence the emergence of exploratory societies to discover new places and implement colonial policies whose objective centred on the domination of territories (JUNQUEIRA, 2003, p. 58; MORAES, 1989, p. 59).

For Capel, the Royal Geographical Society:

> News about discoveries and exploration of new territories, detailed descriptions of them and their cartography were of particular interest. However, this cartographic activity was already carried out by the Ordnance Survey officers, and the Society was often unable to do anything other than report on and comment favourably on the maps that were published. On the other hand, it was very active in drawing up and disseminating rules to guide the work of explorers (1981, p. 177).

It also states that:

> The objectives were much broader than what today is understood as geographical and included not only the organisation of explorations and the promotion of trade, but also

> the creation of meteorological stations, the carrying out of astronomical observations, ethnographic studies (CAPEL, 1981, p. 186).

For Johnston,

> The first of these paradigms was brought into the modern period from the classical period, as exploration was the most important activity recognised as geographical for most of the 19th century. The collection and classification of information about "unknown" parts of the Earth (unknown to Western Europeans and North Americans) was carried out by geographers. Many of their expeditions were financed by the geographical societies that were founded during that century; these, in turn, obtained money from commercial and financial sources.
>
> philanthropic, since the information they accumulated was of great value to the mercantile world. Whilst supporting and sponsoring exploratory expeditions, the geographical societies also played an important role in education. Their conference meetings provided the general public with the opportunity to see and hear about new discoveries, and their directors worked firmly to establish the teaching of geography in schools and universities (1986, p. 57-58).

According to Andrade (1992, p. 49-50),

> Scientists sought to accumulate empirical knowledge and formulate theory; the governments of the countries most committed to colonial expansion, such as England, Prussia, and later Germany, Russia, etc. encouraged the formation of geographical societies that sponsored scientific expeditions into the interior of Africa, Asia and South America in search of resources that could be exploited.

Professor Broek says that "all geographers have in common a curiosity about places, and by 'place' we mean a piece of land and the human group that occupies it" (1972, p. 42). This justifies British imperialism in the territories of Africa, Australia and Oceania.

According to Moraes (1989, p. 59),

> The best known authors of this geopolitical current were Kjelen, Mackinder and Haushofen. The first, a Swede, was the originator of the label Geopolitics. The second, a British admiral, brought the discussion to the level of the general staff, dealing with issues such as the domination of maritime routes, a country's areas of influence and international relations. Halford Mackinder, whose main work is entitled THE GEOGRAPHICAL PIVOT OF HISTORY, developed a curious theory about "pivot areas", which would be the heart of a given territory; for him, whoever dominated it would dominate the whole territory.

According to Andrade (1992, p. 76), "if the English gave greater pragmatic importance to Geography for external use, they also knew how to do it for internal use". England, cornered by Germany, was exposed to bombing raids, razing cities to the ground and causing food shortages as a result of World War II. Then geographer Dudley Stamp (a professor at the London School of Social Sciences) appeared, conducting regional planning for better territorial organisation (ANDRADE, 1992, p. 76-77).

From this perspective, fieldwork is used to recognise and subsequently dominate places through geopolitical strategies and the exploitation of geographical societies. Thus, we can affirm that the colonialism and militarism of the British are very strong marks given to the issue of the resource called fieldwork, and that many European nations embodied it as geopolitical presuppositions.

Moving on, we'll look at the treatment given to fieldwork by the Soviet school of geography.

1.2.4. The Soviet School of Geography

Soviet geography received a strong German influence and aimed to recognise the geography of its territory, as it is gigantic in size with disparate physiographic features.

Capel confirms that "the influence of the work of Humboldt and Ritter was of great importance in the development of Russian natural sciences" (1981, p. 163-164). As a result, with harsh climatic conditions and difficulties in setting up agricultural systems, the Soviets concentrated their efforts on climatic and pedological research.

According to Andrade (1992, p. 58-59), the geotopographical exploration of the Arctic lands (Finland, Siberia and Manchuria), carried out by Kropotkin (an officer in the Russian Army, 19th century), led the Soviets to introduce the branch of glaciology (glacial and periglacial erosion) and the study of permafrost soils to the geographical field.

Still,

> Very concerned about the education and training of young people and admiring the principles of Pestalozzi, Kropotkin condemned the way teaching was done in general, and the teaching of Geography in particular, which, because it was very theoretical and rich in nomenclature, didn't arouse the interest of students; he thought that teachers should use travel books, descriptions, habits and customs of the various countries to arouse the interest of students; take them to the countryside so that, through observation, they could better understand the forms of relief, the structure of the rural and urban environment, in short, the landscape (a term he often used) and, from there, they could better understand and take an interest in the didactic texts about areas that could not be directly contacted (ANDRADE, 1992, p. 61). 61).

It's worth noting that Kropotkin (an anarchist) wrote the chapters on the Physical Geography of Russia and the Far East in the NEW UNIVERSAL GEOGRAPHY by Élisée Reclus (also an anarchist). He was also an in-depth researcher into the physical geography of Siberia.

According to Capel,

> In Russia, during the 18th century, we went from simply collecting information from exploratory journeys, to collecting empirical material and developing geographical science. The interest in geographical issues in 18th century Russia is evident in the fact that it was the first European country to have a translation, in 1718, of Varenio's work, which was later reissued in 1790. From the beginning of that century, Peter I stimulated the realisation of cartographic and atlas work, a task in which the Geography Department created at the Academy of Sciences in St Petersburg played an important role. The most important achievement was the Atlas of Rusia of 1745, directed by Joseph Delisle, which was the first complete collection of maps of Rusia to be produced in a scientific manner and included a number of maps.
>
> on 19 regions, accompanied by data on population and economic activity (1981, p. 159).

Keep scoring,

> In the origins of Russian Geography, the encyclopaedic figure of Mijail V. Lomonosov (1711-1765), chemist, courtier, physicist and poet, who was also director of the Geography Department of the Academy from 1758, is always cited. From there, he was concerned with organising the collection of information through questionnaires sent to the whole of Russia in 1761, carrying out cartographic work (including physical and economic information), and drawing up a general project for land and sea expeditions.
> Soviet geographers attribute to Lomonosov (his 1763 work THE FUNDAMENTALS OF METALLURGY AND MINING) the priority of having established causal

relationships between the different natural phenomena on the earth's surface, using these observations as the basis for geographical systematisation CAPEL, 1981, p. 160).

He goes on to say that "Russian Geography was forged on expeditions organised to study the resources of an immense and little-known country" (CAPEL, 1981, p. 161). The pedological work of Vasilli V. Dokouchayev (1846-1903) in the Soviet lands of Ukraine and Gorki should be highlighted. According to Lepsch,

> In 1877, the Russian naturalist Dokouchayev took part in a commission to study the effects of the catastrophic drought that had occurred that year on the plains of Ukraine. On this commission he had the opportunity to study the region's soils in detail. Years later, he was asked to lead similar work in the forests of the Gorki region, east of Moscow and with a colder, wetter climate. Comparing the lands of these two regions, he realised that those of Ukraine were quite different from those of Gorki and concluded that this diversity was mainly due to differences in climate. He also found that in both regions, the soils were made up of a succession of horizontal layers that began at the surface and ended at the underlying rocks. He recognised and interpreted these layers as resulting from the joint action of various factors that gave rise to the soil, including the climate, and found that each type of soil could be characterised by describing them in detail (2002, p. 6).

And then:

> Dokouchayev was thus able to lay the foundations of soil science. This was not only because he had the opportunity to travel and study an extensive and diverse territory, using his skills, energy and team spirit there, but also because of the "challenge" proposed by his government to understand and improve the productivity of the famous "dark lands" (in Russian chernozem) of the peasant plains, where drought and famine periodically prowled. Using his talented work, with the attitudes and habits of modern science, Dokouchayev recognised soil as a dynamic and naturally organised body that can be studied on its own, just like rocks, plants and animals (LEPSCH, 2002, p. 7).

According to Andrade (1992, p. 80),

> The Soviets were the pioneers in planning the economy and also the pioneers in realising that the importance of geography was not only cultural, academic and political, but also that it could be applied in the field.
>
> territorial planning, this has opened up broad prospects for innovation in the work of geographers.

Moraes (1998, p. 50) states that:

> It was perhaps in Russia that the ideas of Humboldt and Ritter were most literally applied, by authors such as Mushketov, Dokouchaiev and Woiekov. Although Russian geography carried out vigorous fieldwork, it was undoubtedly in Germany that the methodological discussion remained heated.

Thus we have fieldwork in Russia, initially in an exploratory way to properly recognise its territory, and later in a planned way. We can therefore see that fieldwork in the former Soviet Union served effectively to overcome the geographical barriers of the polar climate and glacial areas, and also worked to implement the planning system in its former republics.

Next up is the North American school and its relationship with fieldwork.

1.2.5. The North American School of Geography

The North American School developed in the second half of the 19th century, with the Swiss geographers Arnold Guyot and Louis Agassiz gaining prominence. They settled in the United States and developed studies in Regional Geography and Geomorphology along German lines (ANDRADE, 1992, p. 77). As a result of the work of these geographers, the physical aspects of North American space were boosted by field research carried out by J. W. Powell, in the field of Geomorphology, studying the West of the United States, and by W. M. Davis, a great theoriser of Geomorphology, who established the Theory of the Erosion Cycle (ANDRADE, 1991, p. 77).

Professor Broek reports that:

> Major John Wesley Powell (1834-1902) explored the lands of the West and was a pioneer in describing and explaining land forms. His concern with the practical aspects of colonisation took him far beyond the usual tasks of a geomorphologist. He clearly saw the risks faced by settlers as they entered dry lands. Unless irrigation was possible, the farms would have to be considerably larger than had been provided for in the 1862 Act. Powell therefore surveyed land forms and water resources and proposed measures to ensure the most efficient use of land (1972, p. 28).

He also highlights the work of George Perkins Mars (1801-1882), who had a deep admiration and concern for the conservation of natural resources. Mars referred to the new school of geography led by Humboldt and Ritter and concluded that man was destroying natural resources.

their habitat through merciless exploitation. Mars' work, "MAN AND NATURE, OR PHYSICAL GEOGRAPHY AS MODIFIED BY HUMAN ACTION", from 1864, and later modified to "THE EARTH AS MODIFIED BY HUMAN ACTION", in 1874, is considered to be the origin of the American conservation movement (BROEK, 1972, p. 29).

Also noteworthy is the great contribution of Dean W. M. Davis at the beginning of the 20th century, which was: "a description of the evolution of land forms, classifying them into young, mature and ancient reliefs" (BROEK, 1972, p. 31).

We also have, in this scenario, the geographers inspired by Ratzel, the professor Elen Semple and Professor Ellswort Huntington who continued the determinist theories throughout America (JOHNSTON, 1986, p. 60-61).

According to Sanfanna Neto,

> Huntington (1915), from the beginning of the 20th century, became well known for his polemical conceptions of the influence of climates on the characteristics of peoples. Reproducing the old prejudices about "the tropics" and the "superior" nature of the temperate world, the author states, in his most important work "CIVILISATION AND CLIMATE", that the tropical world does not favour economic development (2004, p. 63).

In Andrade's opinion, Elen Semple was an extremist of the determinist theory, "which served to legitimise American expansion to the West, decimating the indigenous tribes, and to the South, conquering more than half of Mexican territory" and also "as men of temperate climate and white race, Anglo-Saxons, the Americans of New England considered themselves superior to Indians and Latinos and could expand their domains over the lands they conquered" (1992, p. 78).

Far from this radical approach, we have Carl Sauer's contribution in the cultural-historical field.

According to Pedone,

> In the early days of Cultural Geography, Sauer emphasised the importance of fieldwork, when he said that "Geography was first and foremost knowledge acquired through observation, which one then orders through reflection and a new examination of the things one has seen, and from what one has experienced through direct contact comes comparison and synthesis. In other words, whenever possible, the geographer's main training should consist of fieldwork". What's more, he argued that fieldwork shouldn't be prepared, because if the aim was to record previously designated categories, an exploration wouldn't be fully utilised (2000, not paginated).

It also states that,

> In the 1970s, the US abandoned the Sauerian view and fieldwork to begin analysing the everyday practices that participate in the social construction of space. From this perspective, there is also a critique of ethnographic work carried out with a deep-rooted Eurocentric tradition (PEDONE, 2000, not paginated).

Nast (1994 apud Pedone, 2000, not paginated) points out that:

> The "field" in fieldwork is generally treated as a physical assignment, a tendency that comes from Human Geography and its contributions from Physical Geography around attempts to map the terrain, which have often been related to the interests of the organisations, whether academic or governmental, that provided the resources.

On the other hand, we find the contributions of Richard Hartshorne. From the perspective of Sant'anna Neto,

> It wasn't until the early 1930s that an eminently North American geography flourished, when Richard Hartshorne, proposing a re-reading of the work of the German geographer Alfred Hettner, produced between 1890 and 1910, took up a theoretical analysis that sought a third way of interpreting geography, overcoming the dichotomy of Determinism and Possibilism (2004, p. 100).

Moving on,

> Due to its territorial characteristics marked by a varied diversity of landscapes and the philosophical and scientific influences of the Anglo-Saxon schools, among other elements in the formation of this American nation, the United States would become the mecca of pragmatism, where modelling and the systemic approach to planning would find fertile ground for development, via quantification in the production of diagnoses (SANT'ANNA NETO, 2004, p. 101).

Following the Hartshornean perspective, we have Leslie Curry, a scholar of the US economic agroclimatology. For Sant'anna Neto,

> Leslie Curry was one of the most important scholars of Applied Climatology who, following the conceptions of Hartshorne and his Idiographic Geography, developed a perspective of geographical analysis of climate in which the organisation of agricultural space should necessarily start from a conception of climatic attributes, not as determinants, but as an input in natural and production processes. In this way, both global radiation and the main elements of the climate would come to be seen as economic agents and, therefore, players in agricultural production and parameters of its profitability (2004, p. 101-102).

Like this,

> The power of econometrics in economic models led to the need to mathematise geography, which was increasingly focused on economics. Hence the emergence of "Regional Science" in the United States and the new theoretical proposal in North American Geography, including Fred Schaefer's impactful article "Exceptionalism in Geography" (1953) - Hartshorne's ideas on the nature of Geography were vehemently opposed by the young writer (who died when the article was published). The decisive importance of economics, the need to use mathematical language and the pursuit of universal laws - basic conditioning factors
>
> to scientific status - the subsequent "revolutions", known as quantitative and theoretical, were born (MONTEIRO, 2002, p. 14).

Broek states that:

> In geography, as in all the other social sciences, there is much debate about the need to develop quantitative methods. The terminology is unfortunate, because it gives the impression that Geography wasn't interested in the exact measurement of quantities until a few years ago. This is certainly not true; geographers have always insisted on measuring distances, elevations, populations, goods, etc. In reality, the new reform movement is urging geographers to strengthen the scientific content of their discipline by developing more theoretical concepts and proving them through more refined statistical-mathematical procedures (1972, p. 84).

In this way, fieldwork was supported by the postulates of Carl Sauer, Powell, Davis and Mars, since the other followers of this school used statistical-mathematical equations and the office/laboratory to understand the geographical reality of places. However, we note that in the case of the United States, fieldwork in a very specialised way fulfilled the role of exploring the territory and also as a tool for studying natural and social resources.

We understand that exploring and studying territories with their natural and social components are extremely important mechanisms for advancing knowledge in geographic science.

Continuing our studies, we will see how fieldwork is approached in the field of Critical Geography.

1.2.6. The school of Critical Geography

We can see that the historical background to the formation of Critical Geography comes from the Libertarian and/or Anarchist Geography of Élisée Reclus (French) and Pietr Kropotkin (Russian), geographers who lived at the end of the 19th century and the beginning of the 20th century.

According to Andrade, Reclus pointed out that:

> Geographers should make an analysis based on the following principles: that society is divided into social classes as a result of the ways in which the means of production are appropriated; that this class difference causes a struggle between the dominated classes who aspire to better fortune and the dominant classes who don't want to lose control of power and wealth; and finally, that there is a tendency towards the progressive improvement of man (1992, p. 56).

The geographer Élisée Reclus published the works "THE LAND" (1869), the "NEW UNIVERSAL GEOGRAPHY" (1875/1892) and "MAN AND THE EARTH" (1905/1908). In his works, he acted in a meticulous, detailed, descriptive manner, emphasised cartographic illustration and did not make a geographical dichotomy, seeking to clearly punctuate physical and human relationships and interactions (ANDRADE, 1992, p. 57).

According to Andrade, Kropotkin,

> He remained faithful to positivism and admitted that dialectics, as defended by Marx and Engels, could not make a positive contribution to the development of the sciences. He was very concerned about education and the role to be played by geography in the educational process.
> His commitment to naturalism led him to admit a certain physical-natural and ethnic determinism in the evolution of peoples, although he condemned racial discrimination, which was very pronounced among Europeans (1992, p. 59-60).

There is also, on the other hand, the proposal of Active Geography, by geographers Pierre George, Yves Lacoste, Bernard Kayser and R. Guglielmo who opposed statistical-mathematical pragmatism. Thus, Moraes says that Active Geography was the first manifestation of a critical character:

> His proposal was to carry out a type of analysis that would uncover the contradictions of the capitalist mode of production in the various regional frameworks. It was thus a geography that denounced unjust and contradictory spatial realities. It was a question of explaining regions, showing not only their forms and functionality, but also the social contradictions contained therein: poverty, malnutrition, slums, in short, the living conditions of a section of the population that did not appear in traditional ecologically inspired analyses (1998, p. 118).

They still say that:

> The unity of Critical Geography is manifested in the stance of opposition to a contradictory and unjust social and spatial reality, making geographical knowledge a weapon to combat the existing situation. Critical geographers, in their different orientations, take the popular perspective, that of transforming the social order. They seek a more generous geography and a fairer space that is organised in the interests of people (MORAES, 1998, p. 126-127).

In Rodrigues (2008, p. 123-124), Monteiro (2002, p. 27), Moraes (1989, p. 122), Moreira (1994, p. 50-52) and Andrade (1992, p. 123-128), the earthquake proclaimed by critical geographers in relation to Anglo-American statistical-mathematical thinking echoed a picture of changes to the status quo, supported by texts published in the journals ANTIPODE (1969) and HERODOTE (1976).

Yves Lacoste's 1989 work, "GEOGRAPHY SERVES, FIRST OF ALL, TO WAGE WAR", is also worth mentioning, pointing to two geographies: a) the Geography of the Major States and b) the Geography of Teachers (JUNQUEIRA, 2003, p. 70; MORAES, 1989, p. 114; RODRIGUES, 2008, p. 124). The Geography of the Major States is strongly linked to the geopolitical factor, whose focus is based on expansionism and the global market (MORAES, 1989, p. 114; JUNQUEIRA, 2003, p. 70; RODRIGUES, 2008, p. 124-125). Teachers' geography is fuelled by mnemonics, also known as "geography by rote", with the backdrop of hiding socio-economic events and the posture of government power (MORAES, 1989, p. 114; JUNQUEIRA, 2003, p. 70; RODRIGUES, 2008, p. 124-125).

This critical approach is based on the Marxist approach, where
Moreira (1994, p. 52), emphasises:

> What distinguishes men are not the natural elements, but their economic and social conditions of existence. In the deserts of the Sahara, as in the tropics of Brazil or the conifers of southern Chile, what we have are men living under diametrically unequal economic and social frameworks: the fortunate and the hungry. History, that is the substratum of Geography.

Once again, in the face of the Marxist "tsunami", we are left with the teachers' ideas Mamigonian (2008, p. 29) and Andrade (1999, p. 18), that the Critical Current vilified fieldwork and Physical Geography. This situation generated a strong seismicity between critical geographers and physical geographers, culminating in the emergence of isolated geographical events.

We can infer that this conflict is extremely damaging, because we need to remove the rifts, heal them and prioritise the link between geographers. From then on, fieldwork will fulfil its scientific and pedagogical task.

Next, we have to look at the approach that the Brazilian school of Geography gives to fieldwork.

1.2.7. The Brazilian School of Geography

With regard to Brazilian Geography, there is a wide range of views that point to types of periodisation. Monteiro (1980, p. 9-33) attempts to periodise the evolution of geographical research in Brazil from 1934 onwards: "a) The implantation of Scientific Geography (1934-1948); b) The Agebean Crusade for the Dissemination of Geography (1948-1968); c) On the Road to Affirmation: 1ª . Season (1956-1968); d) On the Road to Affirmation: 2ª . Season (1968-1977)".

Later, Monteiro (2002, p. 1-36) presents another periodisation:

> Part I - The Great Evolutionary Stages: 1- The Beginning of the Century (1900-1935) - Preparation for Scientific Geography; 2- Dawn of Scientific Geography (19351956) - 1st. Moment (1935-1948), 2nd Moment (1949-1956) and 3rd. On the Road to Affirmation (1957-1968); Part II - The Crossing of the Great Historical Crisis: 1- The Sill of the Post-Modern and the Great Mutations (1968-1973), 2- The Entry into the Post-Modern (1973-1984) and 3- The End of the 20th Century and the Multiple Uncertainties at the Beginning of the 21st Century (1984-2001).

Andrade (1999, p. 9-13) also proposes a periodisation, or rather, presents a trajectory of Brazilian Geography in three major periods: "The Colonial, the Imperial/First Republic and the Modern".

For the state of Goiás, Professor Gomes (1999, p.13-19) considers the following periods: "Historical or Narrative-Descriptive Period (1722-1938), Ibegean or Institutional Period (1938-1960) and the Academic Period (1960-1996)".

We would like to point out that there are other periodisations of geography, but we prefer to stick with these for the sake of illustration. From what the teachers have said above, I can see that prioritising simplicity is not a sign of poverty and isolation.

For this reason, we have taken the risk of presenting a periodisation, which at first has a scholarly bias, but which gives us significant help in understanding the steps of Geographical Science on Brazilian soil.

Here, without ado, we present my periodisation proposal for Brazilian Geography: NON-CATEDRATIC STAGE and CATEDRATIC STAGE.

The Non-Cathedral Stage refers to the period from 1500 to 1900, a phase that involves exploration, incursions and geographical descriptions by colonial chroniclers, naturalists, Jesuits, bandeirantes, the military and others. This period shows that Brazil's various routes, locations, physiography and biodiversity were wide open to the Eurocentric gaze.

We believe that the post-1900s Catedrático Stadium began with the keen eye of Delgado de Carvalho, when he developed his works, School Geography and also the Higher Free Course in Geography. In this period, after Delgado de Carvalho, we come across the universities in São Paulo (USP - University of São Paulo, 1934) and Rio de Janeiro (UFRJ - Federal University of Rio de Janeiro, 1935), the AGB - Association of Brazilian Geographers (1934) and the IBGE - Brazilian Institute of Geography and Statistics (1938), the XVIII International Congress of Geography of the UGI - International Geographical Union (Rio de Janeiro, 1956), producing a geography with a French flavour, later with an Anglo-American flavour and then with Marxist, environmental, cultural and post-modern biases.

We can see that fieldwork in Brazilian Geography, depending on the Geographical Current, is sometimes valued and sometimes vilified. At this point, we need to highlight and reflect on the fact that regardless of the Geographical Current, "fieldwork is a historical instrument of analysis in Geography" (MATHEUS, 2007, p. 135) and "fieldwork can be considered an instrument of knowledge of geographical reality" (CARVALHO et al, 2007, p. 1410).

Thus, we can say that fieldwork itself is the unique moment of contact with reality itself. What Read (1966, p. 19) proclaims via geologist Nicolas Desmarest (1725-1815), "GO and SEE".

This reality can be seen in the streets, factories, streams, parks, villages, neighbourhoods, plantations, agro-industries, asylums, quarries, hillsides, etc. The important thing to bear in mind is that each context has its own characteristics and peculiarities, which contribute satisfactorily to the development of the fieldwork.

In short, we can say that the universe to be explored by fieldwork is gigantic and any location can be investigated as an object of geographical study. In this way, we can also count on the help of the teacher in defining and listing the characteristics of the location chosen for the field study in accordance with the objectives previously set in the school environment.

Next, we'll look at the emergence of Geography on Brazilian soil and how fieldwork has been used from the colonial phase to the present day.

1.3 The emergence of Geography on Brazilian soil

The historical antecedents of Brazilian Geography refer to the stages of the non-geographers during the Colonial and Imperial periods and the geographers of the First Republic and the Modern Phase, which began after 1930 (ANDRADE, 1999, p. 9).

On Brazilian soil, from the 16th century until the end of the 18th century, academic-scientific activities were scarce. With the arrival of the Lusitanian Crown in Brazil in 1808, there was a ban on foreigners entering the country, as the aim was to safeguard its natural wealth (ANDRADE, 2006, p. 36).

Miranda (2008, p. 1-2) gives the following reasons:

> The 18th century saw a resumption of European expansion, which had advanced timidly during the 17th century. In Portuguese America, the discovery of gold mines led to an increase in population in the interior of the colony. This was not an isolated case. Between 1740 and 1790, Spanish America doubled its effective exploration area. In North America, the Franco-British dispute intensified and fuelled the growth of the European presence. The same happened in India from 1740 onwards.
>
> By 1780, around 30 per cent of the world's population was under the control of the European powers. The new forms of relationship between the latter and the rest of the world tended to intensify and broaden trade flows. Studying the origins of the Industrial Revolution in England, I note how the "growing and increasingly rapid flow of overseas trade" stimulated manufacturing, boosted the market for cheap products such as coffee, tea, cotton and tobacco, as well as expanding the installation of production systems in the colonies, especially plantations. In fact, the economic performance and political stability of the European states were increasingly dependent on their relations with America, Africa and Asia.

In this situation, it was necessary to learn more about the interior of the other continents.

In Brazil, at the end of the 18th century and in the first half of the 19th century, journeys of exploration were no longer the privilege of bandeirantes, chroniclers, Jesuits and the military, but were carried out by European naturalists at the service of scientific academies and museums. The naturalist travel literature on Brazil gained momentum from 1810 onwards, with the policy of opening up the ports and, consequently, the entry of foreigners into Brazil, imbued with the objectives of trade, science, diplomacy, adventure, the arts and militarism. Foreigners from different professions made incursions and descriptions of Brazilian lands everywhere, both near and far.

In the context described above, according to Miranda:

> The great international scientific expeditions began. Seeking to make more intense, rapid and effective use of the human and material resources offered by the various parts of the planet, scientists produced knowledge capable of identifying and evaluating them, offering parameters for rethinking relations between Europe and the other continents. In the instructions for scientific journeys of the European academies, for example, there was a concern to collect and acclimatise plants from distant places, activities that would

be useful for the trade of the great powers or, as happened with the potato, that would help solve the problem of hunger among Europe's poor. The journey of La Condamine and his companions in the 1730s was the starting point for opening up South America to the scientific community. Since then, naturalists have travelled to distant regions to catalogue and classify plants, animals, geographical features and human types (2008, p. 2).

The journeys of the naturalists historically mark the establishment of science in the country. The arrival of the Lusitanian Court inaugurated a period of scientific fertility. Dom Pedro II, known at the time as the "patron of science", was very attached to European scientific theories/theses and constantly advertised the country to European communities as a favourable environment for new academic research (ANDRADE, 2006, p. 50).

Thus, from Miranda's perspective:

> The work of these travellers has many similarities. They were all naturalists (Saint-Hilaire and Martius, botanists, Spix, a zoologist) and had close relationships with important academies of science in their countries of origin (Paris and the United States).
>
> Munich). The Germans came on an official mission and were sent by the king of Bavaria, Maximilian José I, in the entourage of the Austrian archduchess who was to marry Dom Pedro I in Rio de Janeiro; the Frenchman arrived in Brazil accompanying his nation's ambassador. The Portuguese Crown seems to have taken a favourable view of the scientists' presence, as the local authorities, according to the travellers' own accounts, offered them letters of recommendation, accommodation and tax exemptions - from tolls, for example - which facilitated their journeys through the interior. Even after returning to Europe, they continued to maintain contact with the Brazilian elite, who were then involved in setting up the independent state. Martius and Saint-Hilaire, for example, were corresponding members of the Brazilian Historical and Geographical Institute (IHGB) - the former even won a competition organised by the institution. Alongside Brazilian intellectuals, they produced scientific knowledge capable of guiding the definitive implementation of the European civilising model in America (2008, p. 3).

And according to Andrade (2006, p. 50), "through the work of the Emperor, Brazil became a living source for the expeditions of these travellers, including Langsdorff, Von Martius, Von Spix, Koster, Louis Agassiz, August Saint-Hilaire, Castelnau, Gardner and Pohl".

Kury (2001, p. 158) explains that Agassiz,

> Like Martius and Spix, he was brought up in an atmosphere impregnated with Naturphilosophie and the scientific works of Goethe. However, Agassiz's two great scientific references were Alexander Von Humboldt and Georges Cuvier. From Humboldt, Agassiz inherited a concern with the geographical distribution of animals and a love of travelling; from Cuvier, the working methods of Comparative Anatomy and beliefs in the fixity of species and the theory of the great cataclysms that revolutionised the planet.

According to Andrade (2006, p. 37),

> The "fashion" was to analyse and describe the "discovered" countries: to investigate "the other". To get to know, "live", who this other was, which exerted a desire and fascination on European intellectuals, mixed with mysticism, exoticisation and scientificity. This motivation can be considered the driving force behind the activities of foreign travellers on Brazilian soil.

The researcher goes on to show that naturalists,

> Von Martius, Saint-Hilaire, Castelnau and Agassiz, influenced by Humboldt, made the difficult decision to see Brazil with "their own eyes". Organising such trips required

months of preparation: defining the itinerary, organising scientific material, provisions, helpers, letters of recommendation from the Brazilian government and local authorities in the regions visited. Most of the funding for these expeditions came from European governments. Those who didn't receive this kind of funding tried to sell the material they collected to museums (ANDRADE, 2006, p. 48).

The same author also emphasises that:

> The interests of Saint-Hilaire, Pohl, Castelnau and Gardner's trips to Brazil in the 19th century were botanical, geomorphological, zoological and anthropological studies and research: getting to know natural resources and man. Their expeditions were constantly accompanied by artists such as Louis de Choris and Thomas Ender. Iconography and travel accounts went hand in hand and had a place of their own. Drawers and painters on expeditions used their work to promote our country abroad (ANDRADE, 2006, p. 49).

Leinz and Amaral (1989, p. 8) point out some details about the work carried out by foreigners in our lands:

> The geological sciences in Brazil began at the end of the 18th century with a work by José Vieira Couto, in which, among various economic problems, mining, already in a phase of decay, is dealt with. At the beginning of the following century, the Andrada brothers stood out: Martim Francisco wrote DIÁRIO DE UMA VIAGEM MINERALÓGICA PELA PROVÍNCIA DE SÃO PAULO in 1805 and his famous brother, José Bonifácio, as well as carrying out various geological studies in the state of São Paulo, described 10 new minerals during his stay in Europe, which is equivalent to almost 1% of all minerals known to date. He was undoubtedly one of the greatest mineralogists of his time and was a close friend of the famous naturalist Alexander Von Humboldt.
>
> After the Portuguese Court moved to Brazil, two mining engineers were brought into the royal service, the Germans W. L. Von Eschwege and F. A. Varnhagen. The former was in charge of geological surveys in Minas Gerais, mainly targeting the gold region, and Varnhagen was in charge of building the Ipanema Ironworks in the state of São Paulo. Von Eschwege's PLUTO BRASILIENSIS, about our mining industry, became famous.
>
> Thanks to the scientific interest of Empress Leopoldina, the naturalists J. E. Pohl, Von Martius and J. B. Spix came here, great contributors to the natural sciences in general and to Brazil. Pohl's great work, REISE IM INNERN VON BRASILIEN, contains a number of geological and lithological observations that are still valid today on the geology of the states of Rio de Janeiro, Minas Gerais, Goiás and Mato Grosso.
>
> In 1883, the famous naturalist Darwin was here and among the various geological observations he made, the most interesting was that he recognised the island of Fernando de Noronha as being volcanic in nature.
>
> In 1875, the Geological Commission of the Empire of Brazil was set up, headed by C. F. Hartt, who had first been a geologist on the THAYER expedition led by Agassiz, the famous Swiss scientist who came here to study the ichthyofauna of the Amazon. Hartt wrote GEOLOGY AND PHYSICAL GEOGRAPHY OF BRAZIL, a work of great value at the time and still consulted today. He was responsible for the arrival of O. A. Derby and J. C. Branner, two great and immortal figures in Brazilian geology. Derby published 174 research papers, covering all sectors of the geological sciences, as well as carrying out cartographic and historical research. Branner published around 60 works on our geology, edited the first geological map of Brazil, studied the island of Fernando de Noronha in detail and is the author of GEOLOGIA ELEMENTAR, an excellent textbook, elucidated with national examples when writing about and explaining the geological phenomena covered.

Professor Troppmair highlights the evolution of biogeographical studies in Brazil during the time of the travelling naturalists:

> Phytogeography studies in our country began with the observations of European travellers who, crossing Brazil, left numerous works. At the beginning of the last century (19th century), Humboldt (1769 to 1859) stood out, who unfortunately only travelled through a small part of the Amazon, but August Saint-Hilaire (1799 to 1853), Johann Batist Von Spix (1781 to 1826) and Carl Friedrich Von Martius (1794 to 1868) left us important botanical works. During his travels around Brazil, Martius reported on "complex" vegetation formations such as rainforest, grasslands and caatingas and

> associated their distribution and physiognomic aspects with the environmental conditions prevailing in the region. The results of his observations are gathered in the 40-volume work FLORA BRASILIENSIS, which is still of great scientific importance today. At the same time Saint-Hilaire visited Brazil, travelling through Minas Gerais and Goiás, where he made observations on primary and secondary vegetation, and on the aspect of plant succession in the face of human interference. The works of these naturalists published in foreign languages (French and German) have recently been published in Portuguese by EDUSP. These three scientists form the tripod that served as the basis for phytogeographical studies in Brazil. Still in the "travelling" period, at the end of the 19th century, but covering smaller areas, we have the work of two Swedes: Peter Wilhelm Lund and Eugenius Warming, who studied the "Lagoa Santa" region in Minas Gerais. Warming's work "Lagoa Santa" laid the foundations for ecological phytogeography in our country. We should also mention the work of João Barbosa Rodrigues, Director of the Rio de Janeiro Botanical Garden, who produced valuable reports on botanical observations, including some aspects of the Amazon and the high altitude Itatiaia area in São Paulo (1989, p. 14).

Thus, we can see from the use of the fieldwork technique the great scientific value practised by researchers and naturalists in the lands of Brazil. These conceptions and studies are homeric, since many researchers today make constant use of these approaches and detailed observations to compose their research in the most diverse corners of the country.

In fact, far from the paraphernalia used in field activities in South America, and exemplifying Brazil in this period, the tools needed by the naturalist-traveller were based on "seeing with their own eyes" (KURY, 1999, p. 46), the physiography of the "happy tropics" (SANTANNA NETO, 2004, p. 14-23).

In Lima's exposition (2003, not paginated),

> Brazilian Geography has a history that has evolved very significantly. Its history is intertwined with the history of Brazil itself, as the records of the first moments of the nation, told by its first narrators, provide a wealth of significant information.

The researcher goes on to say that the "illustrious visitors who were here gave us the necessary elements to have the first ideas of the nature of our territory in its main aspects" (LIMA, 2003, not paginated). He goes on to say that "the journeys of Agassiz and Saint-Hilaire, for example, left extraordinary documentation that should still be studied today with the greatest attention for what they represent" (LIMA, 2003, not paginated).

Sodré (1976, p. 10), says:

> From the 18th century onwards, however, other researchers began to appear, the foreign travellers, whose efforts followed a scientific intention which, in some cases, was referred to as geographical. The works of La Condamine and Saint-Hilaire are well known. In the 19th century, during the second decade of which Saint-Hilaire visited us, scientific journeys intensified and varied in scope: interesting information appeared on Brazil's ethnography, flora, fauna, geology and even climate. However, these were field trips, in which we were just objects, carried out by foreigners, with a disinterested purpose, as far as we were concerned, never with a practical purpose.

Leite (1995, p. 12) makes it clear that "initially, it was considered that the two stars of the 19th century in whose orbit the others revolved were Alexandre Von Humboldt (1769-1859) and Charles Robert Darwin (1809-1882)".

We would like to highlight the strong contribution of Humboldt's geographical work, which influenced the future naturalists and geographers of his time and of today. Humboldt's geographical work was masterful because, even though he didn't set foot on Brazilian soil, his techniques of research, analysis, description, observation and fieldwork are in the veins of geographers who adopt these methodologies.

Moraes (1983 apud Leite, 1995, p. 12) explains that:

> In 1794, Alexandre Von Humboldt devised a PHYSIQUE DU MONDE, comprising all forms of life and their relationship with physical conditions. It is said that this exponent of Physical Geography and Biogeography was, after Napoleon, the most famous man in Europe, through the influence he exerted by coordinating and guiding scientific and artistic endeavours.

He goes on to say

> His orbit of influence expanded through intense correspondence with contemporary naturalists and through his work in international scientific societies when, after his travels in Spanish America, he spent 21 years in Paris, spending his family inheritance with his collaborators, organising the meteorological, oceanographic, climatic, botanical, zoological and geological material he had gathered. It was then that he formulated theories on magnetism, volcanicity, seismology and tectonics (LEITE ,1995, p. 12).

With regard to Brazil in the 19th century, the naturalists, based on the aspect of Humboldt's work, emphasised "the concern with the question of the variety of natural beings and the need to reduce vital forms to a small number of fundamental types. He combined observation, comparison and generalisation, moving from unity to diversity in order to discover laws" (LEITE, 1995, p. 12).

According to Leite (1995, p. 12), "Humboldt's geography was a study of nature that encompassed humanity as an active species". Professor Sodré (1987, p. 34) says that "Humboldt was essentially linked to Physical Geography, which he undoubtedly revolutionised". For Broek (1972, p. 26), "Von Humboldt was inspired by the desire to understand the complex totality of the universe. In this respect, he was the last of the great cosmographers".

It should be noted that:

> As well as being linked to the inspiration and planning of the circumnavigation voyages, seen as an application of the comparative method, which contributed astronomical, geographical and oceanographic data to the layout of the Kosmos, Humboldt also suggested travelling to the interior of the continents, being a mentor or at least an explicit inspiration to most of the naturalists who studied Brazil (MORAES, 1983 apud LEITE, 1995, p. 12).

Sant'anna Neto (2004, p. 41) points out that Humboldt was influenced by the adventures of travellers Captain Cook and Georg Foster around the world, and "these experiences made a decisive contribution to his education, as he acquired a humanist and global vision, as well as an entrepreneurial and expeditionary spirit".

Like this,

> In fact, it can be said that the origin of Humboldt's intellectual and scientific project derives from the interposition of three currents of thought: two of them scientific, Botany and Geognosy, and a third of a philosophical and literary nature, Idealism and German Romanticism" (SANTANNA NETO, 2004, p. 41).

Another interesting aspect was the contributions made by the Societies

Geographical explorations of the continents, which Capel (1981, p. 173-186) mentions:

> From the second half of the 19th century, Geography became a science at the service of the imperialist interests of European countries. Knowledge of colonial countries was a pressing need for European governments, which therefore encouraged not only the carrying out of explorations, an important basis for geographical knowledge, but also the creation of study centres dedicated to researching overseas countries. These institutions were designed to develop a body of knowledge about non-European countries and to train colonial officials: cartography, geography, tropical medicine and ethnography were among the scientific fields that achieved notable benefits, and geography also had certain concerns, methods and traditions.
> This official concern for the studies of colonial countries corresponded to a strong social demand on the part of the bourgeoisie for knowledge of these countries, with a view to commercial exchanges and the dissemination of European industrial production and culture.
> The very existence of Geographical Societies, which played such a decisive role in the creation of a geographical environment and the development of this science, can also be seen as intimately linked to colonial expansion.

The Spanish geographer goes on to emphasise that:

> The appearance of these societies actually followed the rhythm of the expansive policies of European states. In 1788, the African Association for Promoting the Discovery of the Interior Parts of Africa was founded in London, an antecedent of what would later become the Royal Geographical Society of London. In 1821, the Geographical Society of Paris was founded, which led a languid life until 1860 and whose history is not only the history of geographical studies in general, but also the history of France's colonial and economic expansion over the last twenty years. In 1828 the Gesellsshaft fur Erdkunde in Berlin was founded, and in 1830 the Royal Geographical Society in London (CAPEL, 1981, p. 173-186).

He also emphasises that there are,

> Other societies created before the middle of the century were those of Mexico (1833), Frankfurt (1836), the Brazilian Institute of History and Geography (1838) and the Russian Geographical Society of St Peterburg (1845).
> The American Geographical Society of New York was founded in 1852, the Geographical Society of Geneva in 1858, and the Royal Spanish Geographical Society of Madrid in 1876. Societies of this type were also founded in colonial territories, such as the Geographical Society of Bombay, created in 1833 by British officers to stimulate the exploration of Asia, and that of Quebec - the third in America - in 1877 (CAPEL, 1981, p. 173-186).

Gregory (1992, p. 33-34), reverberates:

> Around 1850, the beginnings of Geography were established, including Physical Geography. These beginnings were expressed in the foundation of geographical societies and the creation of professorships at universities. In the early 19th century, many new scientific societies were founded, and the first Geographical Society to be established was that of France, inaugurated in Paris in 1821, quickly followed by Germany in Berlin in 1828 and the Royal Geographical Society in London in 1830. Each society formed tried to have its own characteristic: the Royal Geographical Society was indeed closely linked to explorations, but its journal was also a forum for many enlightening debates.

Andrade (1992, p. 49-50) emphasises that:

> The governments of the countries most committed to colonial expansion, such as England, France, Prussia and, after 1871, Germany, Russia, etc. encouraged the formation of geographical societies that sponsored scientific expeditions to the interior of Africa, Asia and South America, in search of resources that could be exploited. Even in Brazil, in the first decades of the 19th century, numerous European scientists travelled through parts of our territory, writing books about natural conditions and ways of exploiting them. Books analysing the living conditions of the population, contributing to the development of Cultural Anthropology, a science very close to Human Geography.

In this way, we can emphasise that each geographical society used fieldwork as a tool for recognising places and maintaining European hegemonies. We should also point out that the geographical societies created in peripheral and colonial areas maintained the Eurocentric structures of research and knowledge at their core.

For the Brazilian case, Pereira outlines that:

> During the 19th century, initiatives to explore overseas territories took shape in Europe, in the form of military expeditions or study trips organised by scientific and commercial associations. Among these are the geographical societies, formed from the 1820s onwards and especially prolific between 1870 and 1890, at the start of the imperialist era. Acting as centres of scientific exchange and incentives for exploration, they helped to fill gaps in information about the world and provide elements for its reconfiguration on a representational level.
>
> The emergence of societies of this type in regions of the planet identified by the European gaze as objects of exploitation is, in itself, a fact worthy of attention. The coexistence of two of them in the same country, as in Brazil, is another interesting fact that emphasises this aspect. The IHGB - Instituto Histórico e Geográfico do Brasil was founded in 1838 with the mission of laying the foundations for the political, social and territorial identity of the Empire. The SGRJ - Geographical Society of Rio de Janeiro, on the other hand, was organised at a time of crisis for imperial power and the growing integration of Brazil into the world capitalist economy, when Geography had become the most cosmopolitan of all sciences (2005, p. 113).

Field work, monographs, lectures, events and courses developed

both by the "Historical and Geographical Institute of Brazil and by the Geographical Society of Rio de Janeiro and by the Geographical Service of the Army", are basic, and not abject, for National Geography (EVANGELISTA, 2006, not paginated; CARDOSO, 2006, p. 1-8; ZUSMAN, 2001, p. 8-10).

For this case,

> However, he points out that there was another line besides that represented by the IHGB and SGRJ, the line of Physical Geography, in which the geological and geographical commissions played a decisive role, namely the Geological Commission of the Empire of 1875 (closed in 1876), the School of Mines of Ouro Preto (founded in 1876), the Geographical and Geological Commission of São Paulo (founded in 1866), the Geographical and Geological Commission of Minas Gerais (founded in 1892), the Geological and Mineralogical Service (founded in 1907), whose orientation after 1915 was clearly geared towards the economy, as well as the National Museum in Rio de Janeiro, the Goeldi or Paraense Museum. In this Physical Geography, there was a greater concern with geography as a resource, as well as a concern with maps, charts, etc (PEREIRA, 1994 apud EVANGELISTA, 2006, not paginated).

Pereira also points out,

> Later, in a way, the Brazilian Institute of Geography and Statistics (Instituto Brasileiro de Geografia e Estatística - IBGE) brought the two sides together, both the enquiry aspect (reflective, inculcating perceptions about the country and/or Brazilian places), as well as the census character, based on observations about natural resources.
>
> Another moment in Brazilian Geography was given by the Army's Geographical Service, which had the task of mapping the country at a similar time to the IBGE, perhaps because it had more structure for this type of service.
>
> Thus, the Geography of the 19th and 20th centuries emerged as if in the wake of the preoccupations with Geology and Cartography of the time (1994 apud EVANGELISTA, 2006, not paginated).

In Andrade,

> The Geographical Society of Rio de Janeiro and the historical and geographical institutes in the states have also contributed for dozens of years with studies to gather data and information, not only of historical and geographical interest, but also archaeological, anthropological and political, which can be consulted as a source of information in their

magazines (1992, p. 82).

According to Pereira (1994) quoted by Evangelista (2006, not paginated),

> The real development of geographic-scientific studies and research continued to depend on the progress made in the country by research carried out in the field of related sciences. And the whole period, from the creation of the Geological Commission of the Empire (1875) to the establishment of the National Council of Geography (1937), was, in fact, a period of building and assembling the parts that would make up the complex gears that would transmit movement and strength to modern geographical research.

Cardoso (2006, p. 3) emphasises that:

> The SGRJ also formalised the creation of the Free Higher Course in Geography between 1926 and 1927, aimed at updating primary school teachers. The pedagogical commission was made up of three figures later considered to be the great renovators of geography teaching: Everardo Backheuser, a professor at the Polytechnic School, Fernando Raja Gabaglia and Delgado de Carvalho, both professors at the Pedro II College. The proposal consisted of awarding the diploma of "Laureate in Geography and Related Sciences" at the end of two semesters. In fact, this initiative can be considered a true laboratory for the renewal movement that would emerge in the following decade. To give you an idea, here's a list of the proposed subjects and their respective teachers: Physiography, taught by Delgado de Carvalho; Cosmography, by Professor Honório Silvestre; Anthropogeography, by Professor Everardo Backheuser; Statistics, by Professor Caetano de Oliveira; Climatology, by Professor Jorge Amado; Ethnography, by Professor Heloísa Alberto Torres; Ecology, by Professor Abel Pinto; Oceanography, by Professor Roberto Seidl; Modelling in Geography, by Professor Delgado de Carvalho and the course of excursions through the Federal District (now Rio de Janeiro), organised by Professor Everardo Backheuser.

In Monteiro (1980, p. 14),

> In this phase of training new geographers, the contribution of Brazilian geographers already established in the basic heroic and pioneering phase cannot be overlooked: Delgado de Carvalho and Everaldo Backheuser. The former, since 1917, with "Meteorologia do Brasil" (an absolutely astonishing work for the time) and several notable contributions to truly geographical aspects in different approaches. The second, preferably in the field of Political Geography, even included theorising connotations due to German influence.

According to Sant'anna Neto (2004, p. 62),

> Carlos Delgado de Carvalho was one of the most brilliant geographers of his time and perhaps the first Brazilian geographer, responsible for the densest and most complete analysis of Brazil's climate at the beginning of the 20th century. The author of a vast body of work that includes a wide range of geographical themes, Delgado de Carvalho lived most of his life in Europe, publishing his works in French, almost all of which are (still) unpublished in Portuguese, such as "Un centre économique au Brésil" (1908), "Le Brésil Meridional" (1910) and "Climatologie du Brésil" (1916). But his main contribution came with the publication in London of "Météorologie du Brésil" in 1917.
> It is interesting to note that at this time, the end of the 19th century and the beginning of the 20th century, there were no higher education courses in Geography in Brazil, which were only introduced in 1934 at the universities of São Paulo and the then Federal District of Rio de Janeiro.

In Andrade (1977, p. 8),

> In the early years of the 20th century, works of high geographical interest appeared in Brazil, although not methodologically geographical; it was only with Delgado de Carvalho, who was born and trained in France - it should be emphasised that he was not a trained geographer, but had a degree in Political Science - that the implementation of scientific geographical thinking began in the country.

According to Vlach (1988b/1989a) quoted by Melo (2001, p. 35),

> Before we emphasise the importance of Delgado de Carvalho's contribution, it should be noted that in 1905, in his book "Compêndio de Geografia Explanativa", M. Said Ali advanced towards a Modern Geography, presenting an unprecedented contribution to Brazilian Geography: the study of Brazil in five regions (Central Brazil, Northern Brazil, Northeastern Brazil, Eastern Brazil and Southern Brazil). M. Said Ali's innovations were incorporated by Delgado de Carvalho in 1925, when he published the book "Methodologia do Ensino Geográfico".

According to the author,

> Carlos Miguel Delgado de Carvalho, a teacher at Colégio Pedro II, was, if not the greatest, one of the most important theorists in the teaching of geography. Born in Paris and trained in Political Science, he had contact with Modern Geography theorists from France, England and the United States, which set him apart from other textbook authors of the time.
>
> In 1925, he published the book "Methodologia do Ensino Geográphico (Introdução aos Estudos de Geografia Moderna)", with the aim of discussing the ways in which Geography was taught and the errors contained in this teaching, such as the teaching of Mnemonic Geography, in which students had to memorise lists of geographical information, since teachers demanded memorisation and not an understanding of the subject (MELO, 2001, p. 35).

It's worth pointing out at this point that the fieldwork was part of the proposals Delgado de Carvalho's pedagogy:

> He defended the "comparative" method as the method of Modern Geography and promoted the use of "environmental studies" by teachers. On the other hand, he argued that teachers should, whenever possible, apply the content they were working on in the classroom to Brazil. However, he recognised the difficulties of implementing his ideas in classrooms (MELO, 2001, p. 36).

At this point, it is necessary to turn to Professor Balzan (1969, p. 106) to clarify the concept of "Study of the Environment":

> It's easy to conclude that Environmental Studies is, first and foremost, a non-book activity. It begins in the classroom itself, when it is proposed and planned on the basis of a more general problem, and it also ends in the classroom, when the results of visits, interviews, etc. are explored in depth and evaluated.
>
> But it is experience and, more than that, LIVING. It should lead to maturity, and for that to happen, the student needs to come back from the Study of the Environment changed, richer in experiences than when they left; that through it, the student grows as a person.
>
> It's a permanent activity, not just a physical activity, but mainly a mental one, in the sense of elaboration, which appeals to achieved schemes and puts them into practice.
>
> It is a technique of great importance, because it is through it that the student is brought into contact with the living complex, with a significant whole that is the environment itself, where nature and culture interpenetrate. The student synthesises, observes and discovers. However, as well as being a methodological tool, the Study of the Environment is also "an end in itself", given its enormous informative value. Students develop concepts and acquire historical, geographical, artistic, economic, etc. knowledge, which is incorporated into their cultural heritage.
>
> Studying the environment, therefore, doesn't mean contemplating reality. Rather, it means bringing reality into yourself, taking it on.

Thus, we can see that fieldwork has been "ingrained" in geography from primitive communities to the scientific consolidation of Modern Geography, including the non-cathedral (before 1934) and cathedral (after 1934) periods of Brazilian Geography.

It should be noted that the Geographical Service of the Brazilian Army, the Instituto The Historical and Geographical Society of Brazil, the Geographical Society of Rio de Janeiro and the Geological Commissions used fieldwork to discover and demarcate the backlands, mining areas, landscapes and remote communities of continental Brazil.

Thus, in 1934, national geographic thinking began, with a strong presence of French influence, although it struggled with very different conditions, since there was no breadth of knowledge of our reality compared to French knowledge of theirs, in terms of statistical, physiographic, social and other data. And since Geography was only a subject taught in secondary schools and had not yet acquired the status of a chair in the universities to come, it was left aside, making it difficult to develop research (ANDRADE, 1977, 8-9).

Geography only gained professorship status in the 1930s, with the creation of the country's first universities in São Paulo (1934) and Rio de Janeiro (1935).

In Brazil, exactly

> In the field of Geography, institutions such as the University of the Federal District and the University of São Paulo emerged in the 1930s, with specific higher education courses in this discipline, as well as the Brazilian Institute of Geography and Statistics, with the aim of carrying out the 1940 Demographic Census, drawing up the Brazil to the Thousandth Chart and carrying out field research on the various Brazilian regions (ANDRADE, 1999, p. 69).

In 1934, the University of São Paulo (USP) was founded, a fact that became a milestone in the country's cultural history. In this way, the training of Brazilian geography professionals began and, at the same time, the AGB - Association of Brazilian Geographers - was created (PAZERA JR., 1988, p. 34, ANDRADE, 1999, p. 6970; MULLER, 1961, p. 43-44; MONTEIRO, 1980, p. 10; ZUSMAN, 2001, p. 15-16; RUCKERT, 1997, p. 19).

Professor Ab'Sáber (2004, p. 81) points out that USP, founded in 1934, "was given the great task of preparing teachers for secondary schools and, at the same time, directing its students towards geographical field research".

In this context, we would like to take advantage of the interview that Canal Ciência/Ciência Hoje (1992, not paginated) conducted with Professor Aziz Nacib Ab'Sáber, who explains the details of his geographical life, emphasises the importance of fieldwork and makes it clear that the field trip with Professor Monbeig defined his professional life,

> 1- But it was geography that you dedicated yourself to. What was the reason for that choice?
>
> The price of history books and subscriptions to specialised magazines was an obstacle. On my first field trips, I discovered that in geography I could read the landscape and didn't need books. Nor was there a bibliography for the work we had to do. You just needed to be healthy and willing. So I started going out into the countryside and making short trips. As I didn't have a camera, I learnt to draw the landscapes I saw.
>
> 2- Was it as a student that you began to produce scientifically?
>
> The first fieldwork I did, without guidance, was on the geomorphology of the Jaraguá region and its surroundings. Leaving by train from Luz Station in São Paulo, I described the hills until I reached Taipas and then climbed to the peak of Jaraguá. I discovered that there was something else, apart from history, that I liked to do and that was very sporty: travelling by suburban train, which at that time was very cheap. That's how my scientific career began. While I was still a student, I was already doing research and that's why I ended up getting a bit of a mark.

3- What do you mean?

At the time, not all teachers were born researchers. When they went into the field, it was to make a mise-en-scène, because they didn't have much observation ability. They studied the geography of the landscape: the geometry of shapes and the human use of space. There was still no ecological sense. On my excursions I tried to examine landscapes as a whole, but I soon specialised in geomorphology. Rather prematurely, I recognise today.

4- What were your first activities after graduating?

Between 1944 - when I obtained a bachelor's degree and a diploma in geography and history - and 1965, I tried to get to know Brazil, as I didn't have the money for longer journeys and there was no assistance of any kind. I was lucky enough to join the Association of Brazilian Geographers, which met once a year in different parts of Brazil. The association didn't meet in capital cities, only in small towns, and during these meetings we took the opportunity to do field research in the neighbourhood. The society was fundamental in my life because, as well as allowing me to get to know Brazil, it also enabled me to publish short notes on the areas I travelled in its newsletter.

5- Did society pay your expenses?

I paid for it, because I was a student with no resources. I became a board member very early on and when the Boletim Paulista de Geografia was created, Professor Aroldo invited me to join its editorial board. In this way, I was able to make up for my lack of money. I didn't used to go out to pubs and restaurants because I couldn't afford to share expenses. But I was very lucky with my colleagues. One day, Miguel Costa Júnior suggested that we pool our money to visit a faraway place. With little money and the help of the Brasil Central Foundation, we went - Professor Pasquale Petrone, Miguel Costa Júnior and I - to Uberlândia. There, we discovered a citizen who was taking goods to the city of Aragarças (GO). The Aragarças centre was being built by the Brasil Central Foundation on the right bank of the Araguaia River, opposite a very poor little town called Barra do Garças. This trip was fundamental to my career, because I left a region of hills where I had spent my childhood and ended up in Central Brazil, with endless plateaus, savannahs and gallery forests. For the first time I felt the difference between Brazil's morphoclimatic domains. I then started reading the works of travellers like Saint-Hillaire and became very attached to Central Brazil. I wrote a long work on the south-west of Goiás, together with Miguel Costa Júnior. The work, Contribuição para o estudo do sudoeste de Goiás, is published in the annals of the Association of Brazilian Geographers.

6- Was this your first job?

It was my first full-length work. I had previously written about the geomorphology of Jaraguá and its neighbourhood. All my subsequent work stemmed from that trip to Central Brazil and another I made later, via the Association, to the Northeast. On this second trip, when I came down from Campina Grande (PB), after crossing the Borborema Plateau, to the region of Patos (PB), I saw for the first time a dry mountain range, full of elaborate ridges in plunging quartzite structures. From the tip of this mountain range, I entered the high sertão for the first time, which is low, undulating, with extensive caatingas, intermittent rivers and bizarre hills, like sugar loaves, but called inselbergs due to the semi-arid conditions of their surroundings. I immediately realised that I was facing the third domain of Brazilian nature. For many years I dedicated myself to understanding how far those inter-planaltic depressions with mountains and caatingas, dry region soils, men and sertaneja society were projected by the world of the caatinga. In this regard, I published an article in Ciência Hoje entitled "Os sertões - a originalidade da terra" (The backlands - the originality of the land), one of the first comprehensive works on the backlands region.

7- How old were you when you made your first trip?

The trip to the south-west of Goiás was in 1946, when I was 22. The trip to the Northeast took place later, in 1951 or 1952. In the first phase of my career, I tried to understand the topographical compartmentalisation of Brazil. I had already realised that there were three integrated domains of nature - what today we would call morphoclimatic and phytogeographic domains - and three domains of human geography, with very rustic and suffering human-environment relations. My aim was to understand the general topography of the country, because the maps of that time said nothing. There was talk of the Espigão Mestre and we didn't know if it was a ridge or a dividing plateau. Ahead of this region, which lies between the São Francisco Valley and the present-day Brasília region, there was the unknown. Even less was known about the space between this region and the outskirts of the Amazon. I dedicated myself day and night to understanding the general topographical compartmentalisation of Brazil, this complex system that involves uplands (mountains, plateaus) and lowlands (interplanatic depressions and systems of hills and terraces). This was my first concern, which gave substance to my way of perceiving physical and ecological spaces.

8- Did you have any idea where you would end up with your studies?

In 1956, I established a roadmap for studying geomorphology. I initially set out to understand compartmentalisation and the forms that compartments take, what you see. As a geographer, I had to have eyes. And I was taught this from the start by the French masters. So I tried to develop this perception, because without it it's impossible to be a geographer. From 1956 onwards - under the influence of the great geomorphologists and Quaternary geologists who came to Brazil to take part in the 18th International Congress of Geography, held in Rio de Janeiro - I began to take an interest in the superficial structure of the landscape, in other words, I began to interpret it as a document of the recent past, of the physical and ecological history of the Earth. That's when I got closer to ecology and geoecology. I became interested above all in the physiology of the landscape, in what depends on the climate. I wanted to have a dynamic notion of the physiology of the landscape that integrated all its components: water falling, rocks decomposing, soils forming, in short, a subtle chain of events. I established a tripod of studies: compartmentalisation and forms; the surface structure of the landscape; and the dynamics or physiology of the landscape.

9- How did you arrive at the theory of refuges?

This story began when I came into contact with the great German, Belgian, French, Polish and Russian geographers who came to the International Geographical Congress held here in Brazil in 1956. Suddenly a plane full of geographers, authors of the books I was reading, arrived in Brazil. It was a party! They didn't understand why even during dinner I was trying to be close to them. That meeting marked me out. Until then, I hadn't had the chance to go to Europe and see the work of geomorphologists with much more extensive training than mine up close. In 1957, when Jean Tricart, a great field geographer, returned to Brazil, I assisted him on an excursion to Salto, Jundiaí, Sorocaba and Campinas. One day we stopped near a ravine where there was an occurrence of stone lines over older terrain and, just below, crystalline terrain. Until then, stone lines had been an enigma to us Brazilians. I knew that there was a bibliography about them and I also knew that what was being said about them wasn't correct. But nobody knew how else to explain them exactly. Tricart then told me that those stone lines - which had given geographers so much trouble, each interpreting them in their own way - must actually be a remnant of a stony ground from the past. It could be something similar, although we couldn't say for sure, to certain stone formations typical of the north-east of Brazil. That area we were in must have been a stony ground with caatingas or cerrados in the past, according to Tricart's shrewd interpretation. He didn't need to say any more: I was enchanted by what he was telling me and from then on I dedicated myself to studying the stone lines.

10- Have you always worked alone?

Yes. I've rarely done any collaborative work, which many people might interpret as selfishness. But I didn't have it any other way. I set myself apart from my colleagues by having a certain facility for fieldwork. Of course, those who pursued a normal career were jealous. When I wrote my first article on Jaraguá, my friends wanted to publish it in a guild magazine, but some teachers wouldn't accept it. Not because they criticised the work, but because it wasn't on the "agenda", because they couldn't judge whether I was right or not. The same thing happened with the work on Goiás: based on the observations I had made, I came to the conclusion that the Paraná basin was a beautiful example of concentric cuestas with an external front, as exemplified in Emmanuel De Martonne's book. I then wrote a long interpretation paper and presented it in Goiânia. The geographers who knew the region knew I was right, but they still decided to contest it. How mean of those who are just starting out! I was only 22, young, country bumpkin and provincial. Some said I had too much imagination. I suffered deeply from these criticisms. A publication at that point in my life was very important, and I almost fell from the clouds when Professor Aroldo de Azevedo said he would publish my work. He published it in its entirety, with the bibliography I had used to do it and which presented a different view of basin formation. When the work came out, Professor Aroldo received a letter from Mexico that said: "I really liked the work of this citizen with a complicated name. It represents an effort to read and apply knowledge such as I hadn't seen before." That's when my closest friends realised that the nasty reactions were from jealous people. All my life these people had complained about what I did, and almost everything I did was accepted with great indifference by them.

11- How did you join the university's teaching staff?

In 1959 and 1960 I was in Porto Alegre teaching. They were the only two years I was away from USP and even then I had problems. But while I was there, I got to know the Araucaria plateau better and was able to improve my studies on the phytogeographic domains of Brazil.

12- You were also in São José do Rio Preto, weren't you?

Towards the end of my career at USP, I became director of the Institute of Biosciences and Exact Sciences at UNESP in Rio Preto. But before that, I discovered the fourth great domain of Brazilian nature: the Amazon rainforest. At that time I completed my synthesis work on the morphoclimatic and phytogeographic domains of Brazil, in which I tried to integrate all my regional studies. In this work, I imagined that if there were different domains - one very humid and the other dry - there could be no clear dividing line between them. I studied the contact and transition areas and was able to verify their existence in the field. It was the best work I've done. Between two areas A and B there are combinations of species, with a predominance of those that can better assimilate the area of ecological tension. In the case of contact between three areas A, B and C, there are components common to all three, and the species most capable of surviving also predominate. But in this case, something is already beginning to emerge that is neither A nor B nor C. A new region is formed, a "buffer", between A, B and C.

13- Can you give an example?

The babassu forest, between the Amazon and the dry north-east, or the cipó forest, between the caatinga and the "cold forest" in Bahia. This was my main work, which launched me into the biogeographical community. Until then, biologists didn't have a good idea of geological spaces applied to Brazil as a whole. I then cross-referenced this data with the data I had obtained on the occurrences of stone lines and found evidence that some of these areas had, in the Quaternary period, caatingas and cerrados. I was then able to state that the vegetation found by the colonisers - Atlantic forests, savannah, caatingas, Amazonian forest, araucaria, prairie - was not the same as that which had existed previously, in sub-Atlantic times or in some moments of the upper Pleistocene. The arrangement that existed in the past was radically different, as a result of an episode of tropical fragmentation.

14- When was this work done?

The research was carried out between 1958 and 1968. But I'm still working on it today and I need to write up my most recent conclusions. For example, I discovered thick, continuous stone lines in the Cristalino river valley, west of the Araguaia river in southern Pará, where Volkswagen had a property. I was astonished when I discovered them over very shallow ground and peripheral Amazonian forest. I deduced, firstly, that Volkswagen would fail in the endeavour because, when planting in that soil, the stones could be exposed. Secondly, I deduced that the forested area was the result of a forest that, over the course of a few millennia, covered an area that had once been caatinga. The caatinga was expelled, making way for the forest. I'm currently rethinking these ideas and have come to new conclusions. I've discovered stone lines in the fields of Amapá that document that, in the past, they must have been caatinga, and that the stone lines aren't very continuous because sandy soil predominates. When there is no resistant matrix in the stone, it is impossible to form stony ground; a sandy soil and a caatinga grassland are formed, which in the north-east is called "arisco", a word that derives from the old Portuguese form "areusco". These are different sides of the caatinga. The fields of Boa Vista, for example, were areuscos in the past. I'm still working out the meaning of the Cristalino river valley's pedregal in relation to the retreat of the Amazonian forests to the refuges. The idea I'm getting is that there was caatinga in the Brasília region and in this area of the Araguaia; cerrados and some central patches of forest along the equatorial strip. The peripheral refuges would be on the humid plateau fronts, on the edge of the plateaus and on the middle slopes of the Andes, as far as the humid breezes go.

Atlantic could reach. The conclusion seems quite credible to me. If in the south of Pará, between the Serra de Gradaús and the chapadões of Maranhão, east of the Tocantins River - there are loads of detrital material, made up of very shallow stone lines, it's easy to deduce that this region was drier in the recent past, approximately between 20,000 and

13,000 years ago. And, if this is true, the fragments of Amazonian forest that were very far back or refugees on the humid slopes would later serve as genetic banks for the reconstruction of the great Amazonian forest continuum.

15- What is the immediate importance of these conclusions?

From an environmental point of view, this shows that the framework found by the colonisers, formed over the last 12,000 years, was based on refuges. It was the biodiversity of the refuges that meant that, as the climate became wetter in the Amazon and tropical Atlantic Brazil, the islands of humidity began to merge and coalesce into the total space of the Amazon and along the Brazilian Atlantic coast. That's why I advocate the establishment of large biodiversity reserves, not only for the future of life in Brazil, but also to preserve biodiversity worldwide. I'm doing this out of my own conscience, not in response to international pressure. I think it's from this awareness that new proposals for the preservation of biodiversity in the Amazon will emerge. Knowledge of stone lines is also important for preventing impacts related to the scarification or furrowing of the soils where they occur.

16- What's special about your work as a geographer?

In reality, I was a great traveller and an apprentice geographer. At first, I travelled to get to know a bit of everything and then, as a geographer, to get to know it in detail. At first I wanted to have a macro view of Brazil, but when I realised that part of the recent past was in the superficial structure of the landscape, I had to go down and look at the ravines, acting as a surface geologist. In 1956, I was delighted to see foreign geographers interested only in the documents of the superimposed soils seen in the ravines. At that time, geography for me meant looking at the general organisation of the landscape and the projection of men. Looking at ravines was the job of geologists. The integration I made, looking at the ravine and the total space, was very healthy and even more useful for an integrated view of the physical and ecological world than what I learnt from my eventual masters.

In short, Ab'Saber's testimony is remarkable and demonstrates the geographical saga of both teachers and students in their quest to learn more and more about São Paulo. We have even pointed out that in addition to the classroom, field activities supplemented much of the content taught. That's why we say that the fieldwork tool was extremely important for the first classes of geographers at USP and, today, we try to use this didactic resource as a way of learning about anthropic and natural landscapes.

Fieldwork should be implemented more frequently in subjects and be part of the teacher's work. It is worth emphasising that, from a didactic point of view, it makes a significant contribution to deepening content and recognising reality.

Taking advantage of Ab'Sáber's famous statement, we bring up Abreu's (1994 apud Evangelista, 2006, not paginated) forceful assertion that "it would not be an exaggeration to say that it was in working 'in the field', and not in colleges, that the first generation of geographers truly obtained their training".

Evangelista further clarifies that:

> It is worth considering that this was not strictly due to the French methodological line that prioritised regional studies; we must also consider another aspect, which is that geographers at the time were faced with notorious difficulties in carrying out their research, i.e. census data was precarious, the cartographic base was incipient, reports on different parts of the country were not properly catalogued, etc. Thus, fieldwork

represented a means of better scouring this country in order to better systematise information about it (2006, not paginated).

It should be noted that Professor Ab'Sáber was a student on the Geography course at USP in 1940, and from then on he never stopped doing fieldwork in Brazil.

Returning to USP, in its early years, "under French influence" (PAZERA JR., 1988, p. 34), it relied on the work of French professors Pierre Deffontainnes and Pierre Monbeig to teach Geography (ANDRADE, 1992, p. 83; MONTEIRO, 1980, p. 11; RUCKERT, 1997, p. 19). This context allowed for the organisation of the first chair of Geography at the University of São Paulo, which was held by Professor Pierre Deffontaines and later by Pierre Monbeig. Professor Deffontaines went on to teach on the Geography course and, as a related activity, founded the AGB - Association of Brazilian Geographers (1934), with geologist Luís Flores de Morais Rego, historians Rubens Borba de Morais and Caio Prado Júnior and sanitary doctor Geraldo Horácio Paula Souza taking part (ZUSMAN, 2001, p. 17; MULLER, 1961, p. 43-44; AZEVEDO, 2004, p. 66-67).

The minutes founding the AGB on 17 September 1934 contain the following objectives:

> 1- Periodic meetings of the members with a presentation of a subject in Brazilian Geography by one of the members, followed by discussion;
> 2- Organising joint excursions to study an issue;
> 3- Creation of a library specialising in Geography, thanks to the collaboration of members and donations (books, magazines and letters).

Thus, the spirit of fieldwork, which according to Muller (1961, p. 47), can be seen in objective two contained in the Minutes of 17/09/1934,

> As Aroldo de Azevedo so aptly said on the occasion of the First Congress of Geographers, held by the AGB in the city of Ribeirão Preto (São Paulo) in 1954, "WE DO NOT BELONG TO ANY CITY OR STATE: WE BELONG TO THIS GREAT CONTINENT THAT IS BRAZIL. WE ARE PILGRIMS OF THE GOOD NEWS - MODERN GEOGRAPHY; AND IN THIS CULTURAL NOMADISM, WE PITCH OUR TENT WHEREVER WE SEE FIT, WITH AN INTEREST IN RESEARCH IN MIND." It is in this spirit that the venues for the various assemblies have been chosen, which is why Lorena (São Paulo, a small city of less than 20,000 inhabitants at the time) was the first of the fifteen cities that have already hosted AGB assemblies.

The 1938 Statute (the organisation's first regulation) of the Association of Brazilian Geographers, according to Zusman (2001, p. 18), established the following principles:

> 1- Researching and publicising geographical subjects, especially in Brazil;
> 2- In order to achieve its objective, the Association will hold regular meetings of its members, carry out study excursions, maintain a periodical publication and seek, through the foundation of affiliated nuclei or in cooperation with similar organisations, to spread its activities throughout the country;
> 3- The Association may not take part in political or religious demonstrations or deal with any matter that is foreign to its objectives.

Therefore, research, the dissemination of texts via periodicals, excursions and distancing oneself from all social and political conflict are the objectives that would establish the core of the scientific organisation (ZUSMAN, 2001, p. 18).

In an article entitled "A EXPANSÃO ECONÔMICA DE SÃO PAULO E A ASSOCIAÇÃO DOS GEÓGRAFOS BRASILEIROS" (THE ECONOMIC EXPANSION OF SÃO PAULO AND THE ASSOCIATION OF BRAZILIAN GEOGRAPHERS), published in the newspaper "O Estado de São Paulo" (The State of São Paulo) by Luis Flores de Moraes Rego (a founding member of the AGB), there was a strong distortion of the basic principles of the AGB's 1938 statute, which was based on the development of geographical studies in the country, especially in the state of São Paulo and the surrounding regions. As well as being an organisation for the development of São Paulo culture, it would also play an important practical role in the modern evolution of Brazil's economic life (Zusman, 2001, p. 18-19).

This makes it clear that some members of the AGB are closely involved with the politics and economy of the state of São Paulo.

In 1935, Pierre Deffontaines left his activities in São Paulo and moved to Rio de Janeiro, where he continued his work on geography at the recently inaugurated University of the Federal District, now the Federal University of Rio de Janeiro (UFRJ). With the transfer of Deffontaines to Rio de Janeiro, Monbeig took over the teaching of Human Geography and also the direction of the Association of Brazilian Geographers in São Paulo (ZUSMAN, 2001, p. 20; ANDRADE, 1992, p. 83).

Andrade (1992, p. 92), reports that:

> The AGB's great contribution to the development of Brazilian Geography in the period under study stems from the fact that it brought together geographers from different parts of the country to debate themes and issues and to carry out field research together; it disseminated methods and techniques as well as the principles prevailing in the most advanced centres.

For Ruckert (1997, p. 22),

> One of the great lessons of the first phase of the AGB was the prospecting of the national territory through the legendary field work carried out by teams divided into research groups. In this first great golden phase, the contribution of Brazilian geographers was significant in shaping a vision of Brazil and perhaps nationalism.

He also states "the importance of the AGB and its exploratory excursions for the formation of a nationalism that is evidenced, for example, in the life and work of Orlando Valverde" (RUCKERT, 1997, p. 22). We should emphasise that Professor Valverde carried out various works on the agrarian question in our country.

In the Ageban crusade of national diffusion (1948-1956), Professor Monteiro explains that:

> This second segment, although not profoundly different from the previous one, seems to have a special characterisation. The enthusiasm following the reformulation of the AGB (1945) was sparked by the Lorraine Assembly (1946). Above all, joint fieldwork began to motivate and interest more and more geography neophytes (1980, p. 15).

Returning to the question of the replacement of Deffontaines by Monbeig, Andrade (1992, p. 85) comments: "without having the resources of USP, the University of Brazil also had its teaching staff made up of foreign professors, such as Pierre Deffontaines and Francis Ruellan, the former in the area of Human Geography and the latter in Geomorphology".

According to Pazera Jr. (1988, p. 34),

> Over the next two decades, the AGB played a decisive role in training young Brazilian geographers in field research and writing monographs, thanks to its famous "General Assemblies". At these meetings, under the direction of more experienced geographers, team research projects were carried out. French geographical thought dominated the Agebean production at the time, with its keen spirit of observation.

For Andrade (1977); Geiger (1988); Thery and Droulers (1991) cited by Zusman (2001, p. 20), "Pierre Monbeig (1908-1984) is considered by Brazilian and international historiography to be one of the 'founders' of Brazilian Geography".

For the researcher,

> Monbeig became the spokesman for an academic project for the subject that had its pillars in Regional Geography, the concepts of landscape and the geographical complex, and fieldwork. This project is explained in his conference entitled "Didactic Orientation" (1935). In it, Monbeig seeks to encourage the establishment of a chair of Regional Geography, the complementing of teaching and research activities and the carrying out of fieldwork (ZUSMAN, 2001, p. 21).

It also states that:

> Monbeig thus proposes a working method for the geographer: successive approximations which, involving successive syntheses, enable knowledge of reality. However, it would seem that the method so clearly explained was presented to Monbeig as stages in the research process, not as a method in itself. In his explicit discourse, the method is associated with fieldwork, a key component in the construction of geographic knowledge (ZUSMAN, 2001, p. 24).

There's another consideration,

> But it was necessary to remind the old students, and probably teach the new ones, that geography, first and foremost, describes; that the geographer then places the facts, recorded and catalogued, in the cosmos; that he carries out a work of synthesis. His method? No theory will show it better than practice (MONBEIG, 1940b apud ZUSMAN, 2001, p. 24).

In addition to the creation of geography courses in São Paulo (1934) and Rio de Janeiro (1935) and the foundation of the Association of Brazilian Geographers (1934), in 1938 the IBGE - Brazilian Institute of Geography and Statistics - was created during the government of Getúlio Vargas (the so-called Estado Novo period, or rather, the dictatorial regime), with the aim of setting up a structure to survey Brazilian reality by carrying out fieldwork in the various corners of the country (ANDRADE, 1999, p. 52).

According to Monteiro,

> One of the consequences of this event was the hiring of the eminent German geographer Leo Waibel, who had been taken to the United States by the Second World War (more directly, Nazism) and the University of Wisconsin, where he had taught some of the geographers of the National Geographic Council, as a technical assistant to the IBGE (1980, p. 11).

The University of São Paulo and the Brazilian Institute of Geography and Statistics (formerly the Brazilian Council of Geography, Cartography and Statistics) used to send candidates abroad to France and the United States for academic and scientific training

(ANDRADE, 1999, p. 73; MONTEIRO, 1980, p. 11), and so the hiring of Leo Waibel for the aforementioned position was justified.

At the time, the IBGE's contributions were associated with promoting courses, conferences, field research, territorial planning work, geopolitical studies, organising the publications "Boletim Geográfico (1943/1978) and Revista Brasileira de Geografia" and creating a career for geography professionals in the country (ANDRADE, 1992, p. 88-91).

In Rio de Janeiro, we have the duo Francis Ruellan and Leo Waibel, at a time when fieldwork and direct observation of nature was an essential condition for Geography and their influence in this area was excellent (MONTEIRO, 1980, p. 12).

Andrade explains:

> The University of Brazil (formerly the University of the Federal District and now the Federal University of Rio de Janeiro) had great links with the IBGE, since many newly graduated geographers went to work at the Institute, which also used the University's professors to give holiday courses to teachers from various states. Foreign masters who stayed in Brazil for a relatively long time worked simultaneously at both institutions (1992, p. 85).

One detail remained unresolved: the fact that the "Association of Brazilian Geographers, despite its name, was a São Paulo institution for many years" and it was only "in 1944 that the AGB became truly national, after the geographers of São Paulo and Rio de Janeiro began to work together at the General Assembly held in Lorena, São Paulo State" (ANDRADE, 1999, p. 70-71).

In 1945, the first change to the AGB's statutes effectively took place, with regional sections appearing in São Paulo and Rio de Janeiro, followed by Lorena (a municipality in São Paulo) in 1946. From then on, the General Assemblies became annual, with communications, exchanges of experiences and, above all, fieldwork (MONTEIRO, 1980, p. 15; ANDRADE, 1999, p. 72, MULLER, 1961, p. 46-47; ANDRADE, 1994, p. 74-75).

From the assemblies in Lorena (São Paulo state) in the 1940s to Londrina (Paraná state) in the 1960s, with the exception of the administrative assemblies in São Paulo (1945) and Rio de Janeiro (1956), the others were graced by the development of fieldwork (MULLER, 1961, p. 52-54).

With this change in the statute following the Lorena assembly, the Association of Brazilian Geographers became a national organisation, regional sections began to spring up across the country and its scientific bulletins appeared.

Muller (1961, p. 46) informs us that "scholars of the new discipline were noticing the interest in the meetings of the still modest São Paulo group, and wanted something similar to be created in other centres of the country".

Currently, the website of the Association of Brazilian Geographers (2007, not paginated) states that:

The AGB is a non-profit civil organisation that brings together bachelors, professors, researchers and students of Geography. One of our main objectives is to encourage professional organisations, student bodies and community groups to work together to improve democratic institutions and improve the living conditions of the Brazilian people.

The Association of Brazilian Geographers (AGB) was founded by a group of researchers led by French geographer Pierre Deffontaines in São Paulo on 17 September 1934. Since its inception, the AGB has brought together renowned intellectuals, including those from other fields of knowledge, such as Caio Prado Júnior, Luiz Fernando Morais Rego, Fernand Braudel, Rubens Borba de Morais, Claude Levy-Strauss and Pierre Monbeig. This character of plurality and democratic radicalism in its composition is a mark that endures to this day in the organisation's daily life.

From 1944 onwards, the AGB became a national organisation, bringing together members, professionals, students and collaborators from all over Brazil. At this time, the first regional sections were formalised in the states of Rio de Janeiro, Minas Gerais, Paraná, Pernambuco and Bahia. In 1946, the AGB held its first national meeting in Lorena, São Paulo, followed by numerous annual meetings until 1955.

In 1954, the First Brazilian Congress of Geographers was held, the biggest event in Brazilian Geography, held every ten years and now in its sixth edition (Goiânia, 2004). In 1956, in Rio de Janeiro, the AGB organised the XVIII International Congress of Geography of the International Geographical Union (UGI).

Until the early 1970s, the AGB was characterised much more as an association of researchers. From that decade onwards, the growing demands of an ever larger and more complex geographical community led to the organisation of the first National Meeting of Geographers in Presidente Prudente in July 1972. These were difficult times for the organisation of Brazilian society, but the National Geographical Meetings (ENG) have consolidated over the years and today make up the second largest technical-scientific event in Brazil.

These demands and struggles for the democratisation of society led to intense debates during the 3rd National Meeting of Geographers (Fortaleza/Ceará, 1978) where the AGB began a profound renewal of its organisational perspective, making it an association even more integrated with the various struggles for human rights and political and democratic debate in society.

The institutional history of the AGB is integrated with the history of geography and Brazilian geographical thought. Much of the scientific output of Brazilian Geography is published in the Proceedings of its Congresses and Meetings. The AGB is also responsible for the national editions of Terra Livre magazine and the electronic newspaper Notícias da AGB.

The AGB is currently organised through a National Executive Board, with two-year terms, and Local Sections that can cover one or more Brazilian municipalities. The AGB also has representatives on important public policy bodies such as municipal, state and federal councils, as well as seeking permanent dialogue with the bodies that regulate the geography professions (CONFEA and MEC).

According to Andrade (1992, p. 82-83),

> The study and teaching of Brazilian geography at university level, however, was only institutionalised after the 1930s Revolution (Getúlio Vargas' government), when the Faculties of Philosophy, Sciences and Letters were created at the University of São Paulo (1934) and the University of the Federal District (1935, now the Federal University of Rio de Janeiro). Also in the 1930s, the Federal Government created the Brazilian Institute of Geography and Statistics in Rio de Janeiro, with three councils, the Geography Council, the Cartography Council and the Statistics Council, which would be used to develop knowledge of the national territory and to rationalise the policy of collecting statistical data, with an influence on the administration itself. The IBGE was the institution that first recognised the existence of geography professionals, not dedicated to teaching, but "to research, although it did supply professors to numerous universities. Also from 1934 was the foundation of the Association of Brazilian Geographers (AGB), initially organised by Professor Pierre Deffontaines and which for several decades provided notable services to the development of Geography in Brazil.

The geographer also emphasises:

> We can admit that the thinking of the French Classical School dominated Brazilian Geography from the establishment of these institutions until the XVIII International Congress of Geography, held in Rio de Janeiro in 1956; from then on, the influence of masters from other nationalities on Brazilian geographers began to be felt; this influence created great disquiet in geographic circles and gave geographers a greater say in planning at national and regional levels. This transformation cannot be marked with certainty, except to establish a periodisation. This is why we recognise the dominance of classical Lablachian thinking in Brazilian Geography from the early 1930s until 1956, with three work fronts being responsible for spreading and adapting this thinking: the Universities, the IBGE and the AGB (ANDRADE, 1992, p. 83).

In 1956, in the city of Rio de Janeiro, the XVIII International Congress of Geography took place, which represented a watershed in the history of Brazilian Geographical Thought, as it promoted significant changes.

These significant changes are presented by Andrade (1992, p. 87),

> When the 18th International Congress of Geography was held in Rio de Janeiro, a large number of professors and students from this and other universities attended, writing excursion guides, presenting theses and papers and taking part in debates. It can be said that Brazilian Geography was mature and able to participate in the great transformations that were taking place in its nature and methodology.
>
> Among the great benefits brought to Brazilian Geography by the XVIII International Congress of Geography were the various courses given by the great European and North American masters at Brazilian universities. The main one was the High Geographical Studies Course, co-ordinated by Hilgard Sternberg at the University of Brazil for 40 assistant professors from Brazilian universities, in an attempt to give them an overview of the current state of geographical science. In this course, Erwin Rainz taught cartography; Carl Troll taught phytogeography along ecological lines; André Cailleux taught sand and pebble sedimentology, with practical work and the use of statistical methods; Pierre Monbeig taught geomorphology, about the cycle of erosion in humid tropical climates; Pierre Deffontaines taught geography of livestock farming; and Orlando Ribeiro taught historical geography of Portuguese expansion in the world.
>
> At other universities, courses were given, such as Jean Tricart's on morphoclimatic zones, in which he criticised the geomorphology of W. M. Davis, demonstrating that erosion systems are dependent on a zonal distribution, a theory that had been developed by the French master, with a strong dialectical background and under the influence of German masters. These ideas were later developed in epoch-making books.
>
> From the courses given in various states of Brazil, the discussions held at the Congress itself, the conferences held in various locations and the publications distributed during the conclave, Brazilian geographers were led to a greater reflection on the methods, techniques and objectives of Geographical Science and on the nature of Geography and the objectives to be achieved through its use. This paved the way for the gradual abandonment of the classical model and the search for new methods that would naturally depend on the training and convictions of the geographers, as well as the political and administrative evolution of the country.

Monteiro (1980, p. 18) points out that in 1956:

> The XVIII International Congress of Geography, held in Rio de Janeiro (8-18 August), is the milestone in the transition from the formation phase to the affirmation phase, where it is hoped that the fruits of the existence of an active community of research geographers will be harvested.

He goes on to say that "by visiting our country on excursions, before and after the Congress and during the courses, these specialists were able to observe various facts about the intertropical domains and compare them with those of other regions of these domains that they knew" (MONTEIRO, 1980, p. 18).

Evangelista (2004, unpaginated), analysing material from the XVIII International Congress of the International Geographical Union - UGI (Rio de Janeiro, 1956), presents the itinerary of the excursions that had as their objective:

> In order to offer congress participants a better knowledge of the physical and human geography of the different regions of the country, nine excursions were organised, four of which took place before the congress and five after it, so as to give the same participant the opportunity to come into direct contact with the Brazilian land twice.
>
> Among the routes taken, the excursions focused on the Amazon, the Centre-West and the South of the country; in the East, the Northeast, the state of Bahia and more particularly the Southeast of Brazil. In the latter region, the geographers travelled four different routes: from Belo Horizonte to the Rio Doce Valley, the coast of Rio de Janeiro, the Paraíba Valley and the São Paulo region (which included the west of the country and the north of Paraná in order to follow the trajectory of coffee).

> The facilities obtained, whether by federal, state, municipal or autonomous authorities, or private companies asked to collaborate in organising the excursions, allowed the Executive Secretariat to establish reductions in the fees for the same excursions, which could be offered to congress participants at a price significantly lower than those announced in the First Circular. For six of the excursions, the reductions reached 60 per cent of the amounts previously set, and for the others the reductions were between 35 and 50 per cent. The prices included transport, accommodation, meals (drinks included), luggage transport to hotels, as well as a shuttle service to each stop. Temporary insurance for personal accidents was also included.

We can state that the 1956 Congress in Rio de Janeiro acted as a great topographical watershed, as it called into question the pedagogical actions of many geographers. We emphasise that this event touched a nerve and redirected scientific approaches, created new conceptions and showed that there are different perspectives on the object of study. As a result, we see that fieldwork is now being given more careful attention from a methodological point of view, thus avoiding its demise.

After the 1956 Congress, Brazilian geography took the path of the Anglo-American School of Theoretical and Quantitative Geography. During this period, around the second half of the 1960s, the situation showed strong signs of change. É,

After about ten years, the so-called Quantitative School, which originated in the United States and England, arrived in Brazil. In Brazil, Theoretical and Quantitative Geography entered UNESP - Universidade Estadual Paulista "Júlio de Mesquita Filho" in Rio Claro (São Paulo) and in the state of Rio de Janeiro, together with IBGE - Instituto Brasileiro de Geografia e Estatística and, to a certain extent, UFRJ - Universidade Federal do Rio de Janeiro (ANDRADE, 1999, p. 74-75).

According to Andrade (1977, p. 16),

> In any case, the introduction of Quantitative Geography in Brazil, which found its great defenders in the IBGE and the Faculty of Philosophy, Sciences and Letters of Rio Claro (now UNESP), São Paulo, did Geography a great service because, by launching a contestatory movement with great vehemence, it provoked an intensification of geographical studies and called our geographers to reflect more on geographical theory. Hence the great usefulness of the modest but radical Boletim de Geografia Teorética and the more recent and less radical Revista Geografia, which replaced it. In reality, this journal seeks to be a bridge between so-called Theoretical Geography and so-called Traditional Geography, trying to make a synthesis out of thesis and antithesis. However, in its first two issues, it remains very theoretical, a fact that is as much a result of the great productive capacity of the young geographers: Christofoletti, Ceron and others, as well as a certain reserve towards the Journal on the part of the best so-called traditional geographers in Brazil.

Theoretical-Quantitative Geography "developed studies and published numerous works based on quantification, despising and condemning the entire past of geographical knowledge. The Association of Theoretical Geography (founded in 1971), with its Bulletin, was a radical bastion of this movement" (ANDRADE, 1999, p. 75).

According to Pazera Jr. (1988, p. 35),

> However, the big impact came towards the end of 1960. It was the current represented by the English-language authors who brought about the "Quantitative Revolution" here. The term is perhaps an exaggeration today. At the time, these texts by authors who used quantitative methods and showed theoretical concerns had a great impact.

Moreira (1994, p. 46) emphasises the quantitative and theoretical revolution in which "instead of field research, the computer takes its place". According to Moraes (1998, p. 96), "the renewal movement will seek new techniques for analysing geography. From an instrument developed at the time of the field survey, an attempt will be made to move on to remote sensing, satellite images and computers". He also points out that "for the authors affiliated to this current, the geographical theme could be explained entirely with the use of mathematical methods" (MORAES, 1998, p. 102).

In Rodrigues,

> Proponents of the Theoretical and Quantitative Current consider the following principle: if mathematics is the language of the sciences in general, then it must also be the language of geography, because through mathematics it is possible to formulate theories in the field of geography.
>
> Geography and also the use of theories from other sciences. Therefore, great importance is attached to maths, especially statistics, as it can guarantee the accuracy and reliability of the results (2008, p. 110).

It is also emphasised that:

> There has been, and still is, a lot of criticism of geographers working within the context of the Theoretical and Quantitative Current, due to the use of models, the mathematisation of society and for considering the methodological and language uniqueness between the Social and Natural Sciences. Theoretical-Quantitative Geography survives, however, without the heyday of the 1960s and 1970s (RODRIGUES, 2008, p. 111-112).

Andrade (1999, p. 75) therefore points to an important issue:

> Over the years, many quantitative geographers came to realise that the methodology adopted was insufficient and that other paradigms should be used, while their opponents, after the acute phase of the struggle, realised that the use of mathematics and statistics was unacceptable as an end, but was very useful as a means, and could be used without the previous excesses. The struggle subsided from a scientific point of view, but became exacerbated from a political point of view.

Professor Moreira (1994, p. 47) points out another detail:

> But the Viet Cong knew the terrain better than the American computer, and the extraordinary volume of international trade came to a head with irrepressible inflation in the 1970s. Quantitative and Theoretical Geography also went into a tailspin.

According to Moraes,

> This is the meaning of the impoverishment mentioned, which is accompanied by technical and linguistic sophistication. The discourse is essentially poorer, with a richer and more elaborate language. However, the instrumental sophistication conveys a more simplistic content (1998, p. 110).

Professor Evangelista (2006, not paginated) emphasises that in 1970, there was a further change to the AGB statutes, and the practice of fieldwork during meetings was abolished. As a result, the participants gathered in Presidente Prudente (São Paulo) for the First National Meeting of Geographers (1972) no longer carried out field research. In an attempt to unravel what happened in Presidente Prudente, in the case of the disappearance of team fieldwork from the event's programme, we have some geographers' considerations:

"We can highlight the fact that the 1st National Meeting of Geographers, held in Presidente Prudente, was marked by a dispute between quantitative and traditional geographers" (MONTEIRO, 1980, p. 31).

For Evangelista (2006, not paginated),

> It can be seen that there is a process of popularisation of geographers' meetings, so that they are not restricted to just a few people. We seem to be facing a
>
> A symptom of the fact that fieldwork was becoming expensive, or that, at the same time, there were no longer the necessary resources for it to exist to its full potential.

In Andrade,

> The Association regularly and periodically organised general meetings in cities, almost always small in population, where members presented papers to be debated and published in the Annals, if approved by the full members, and carried out field research, of which preliminary reports were written, also to be published. The first scientific meeting was held in Lorena, São Paulo, in 1946, and was followed by annual meetings in various Brazilian cities until 1955. There was no meeting of the association in 1956, due to the 18th International Congress of Geography in Rio de Janeiro. The system continued to operate for successive years, until it became clear that the dominant system could not be maintained and the AGB's statutes were reformed to hold meetings every two years. The unfeasibility of the dominant system in the period under study was due to the growing number of members and the impossibility of fitting them into the various field research teams. The biannual meetings continued until there was greater pressure from the grassroots at the Fortaleza meeting (1978), which led to a new reformulation of the statutes, reformulated at an administrative meeting held in São Paulo in 1979 (1992, p. 92).

Vieira and Pedon (2004, p. 78) comment that:

> The 1972 ENG - National Geography Meeting, held in Presidente Prudente, marked the beginning of the paradigm shift in Brazilian Geography, with the centre of the debate being between Quantitative Geography and Critical Geography, with a Marxist bias, as well as discussing the issues that were occupying space in the media and universities at the time, such as major development projects like the construction of the Transamazonian Highway and colonisation projects in the northern region of Brazil.

It should also be pointed out that until 1970 we had the AGB Assemblies, but as of 1972 saw the National Meetings of Geographers.

Vieira and Pedon emphasise that:

> At the Third National Meeting of Geographers, in 1978, in the city of Fortaleza, the definitive milestone in the rise of Critical Geography to the centre of the debate on geographical science took place, with the presidency of the National AGB being held by a member from Presidente Prudente, Professor Marcos Alegre, who had taken up the post provisionally with the commitment to modify the statute (2004, p. 78).

It follows that:

> At this event, the organisation's new statutes were approved, with the participation of the student member category, with a 50% discount on annual membership fees, and the creation of local sections to replace the regional sections and local nuclei. This was an important moment in what is known as the democratisation of the AGB, the result of a long period of debate about the organisation of the AGB which, according to many critics, still had an elitist character because it did not allow wider access to students (VIEIRA; PEDON, 2004, p. 78).

Professor Antunes says:

> The episode of the Third National Meeting of Geographers in 1978 in Fortaleza actually expressed a process that was already taking shape in society
>
> The search for the guarantee of democratic rights has never really left Brazil. Let's

remember that at the time we were still living under repression imposed by what we might call the dictatorship of the elites under the control of the military. The III ENG was strictly a meeting. Not just in the formal sense of geography professionals, but a meeting of experiences that had been developing all over Brazil, in different places, by different people, from a critical perspective. A meeting that took place at a time when Brazilian society was undergoing major transformations, with the re-emergence of important social agents such as the labour movement and the student movement. This event, which essentially became a classic watershed, reflected the processes related to the dissatisfaction and concerns that were already tenuously shining through about the direction of this science in the country (2004, p. 179).

For Moreira, the near rediscovery of geography began

> When Brazilian geographers gathered in Fortaleza in 1978 for the 3rd National Meeting of the AGB, Brazilian Geography was in a state of ebullition. In the various corners of the country, movements of criticism and renewal were flourishing, spontaneously and without national hegemony. Mutual knowledge leads to an awareness of the concomitant discontent that precipitates the crisis of a science that, from Lacoste's texts, we knew to be worldwide, thus promoting the discovery of motives and favouring the agglutination of ideas (1992, p. 6).

According to Gomes (1999, p. 104),

> At the National Meeting in Fortaleza (1978), the so-called youth power was strengthened and the dualistic tendencies became more acute. This moment in time is considered by many to be the age of populism. Although it was considered, especially by the more conservative members, to be qualitatively low, unscientific within the parameters of geographical science and rather partisan; although it revealed the mark of a certain turbulence and lack of coherence; although the partisan-ideological focus often prevailed over gnoseological and professional quality, etc., we understand that the so-called "turbulent period" was of fundamental importance for the development of Geographical Thought in Brazil.

Andrade shows an interesting picture of the meeting in Fortaleza,

> The decisive shock came at the National Meeting in Fortaleza in 1978, when the community was strongly divided and the students, many of whom were not associated with the AGB, took control of the decision-making process. This was followed by the administrative meeting in São Paulo in 1980, when power was won by student groups supported by some geographers and professors (1999, p. 75).

In the interviews section of the Boletim Paulista de Geografia (2008, p. 21), Professor José Bueno Conti, from the University of São Paulo, gives a very clear account of the 1978 episode in Fortaleza:

> Because of the procedure, which I don't think respected different opinions, it was a kind of steamroller. That was one of the reasons. Another was the introduction of Critical Geography as hegemonic in Brazilian Geography at the time. And I, well, I never thought Critical Geography was important, and everyone can have their own point of view, of course, and can debate, in an academic way, but not there: Critical Geography came in overwhelmingly, in a way that excluded Physical Geography and Cartography, because there was no place in Critical Geography for the study of Appalachian relief, the mid-latitude climate. So they excluded Physical Geography from the geographic debate, and I think that was very serious, because it impoverished, mutilated the
>
> Geography. Then it passed, that wave passed, and today it's not like that anymore. That was the other reason I disagreed with the events in Fortaleza.

In the same Bulletin, also in the interviews section, we have the testimony of Professor Armen Mamigonian, from the University of São Paulo, who says that "another cardinal sin is that since 1980, fieldwork has been disregarded in the AGB in favour of theory, as well as disregard for Physical Geography" and also "in fact, the effervescence of 1978/1983 created a lot of noise and little light" (2008, p. 29).

Moreira (2007, p. 30), "however, the Marxist strand, although hegemonic, is nevertheless a strand. Exclusivised, the Marxist strand gains fame, but effectively does little for itself".

On a keyring,

> The 1970s formally saw the genesis of the movement to renew Brazilian Geography called "Critical Geography", when a process of impetuous criticism of "Traditional Geography" took hold in universities. Authors rooted in "left-wing thinking" aimed to fulfil their political mission in the practice of a Geography that, academically and ideologically, would become an instrument of social liberation and the construction of a new historical path (1996, p. 8).

According to Moraes (1998, p. 116),

> The purpose expressed by Lacoste clearly defines the objectives and stance of Critical Geography. It takes on an explicit political content, which can be seen in his statement, "Geography is a social practice in relation to the earth's surface", or in David Harvey's, "the question of space cannot be a philosophical answer to philosophical problems, but an answer based on social practice"; it can also be seen in Milton Santos' statement, "space is man's home, but it can also be his prison". It can be seen that geographical renewal is now thought of, in terms of theory and practice, as a revolutionary praxis, in the sense that it is not enough to explain the world, but it is necessary to transform it".

Souza says:

> Critical Geography emerged in deep criticism of Classical Geography and Quantitative Geography, challenging dominant thinking and participating in the process of transforming society. In the 1970s and 1980s, Critical Geography reinterpreted the aspects of Geography based on Marxist theory, that is, dialectics and historical materialism. Man, seen as a passive being, is now seen as the main actor in the environment, producing its space (2006, p. 45).

Over the years, from the 1970s onwards, fieldwork and the observation and description of places were seriously criticised by Marxist geographers. These geographers claimed that fieldwork did not take into account social, political and economic issues and was therefore a positivist and inadequate practice. Quantitativists believe that the geotechnological apparatus dispenses with the need to go into the field, because the electronic view of the landscape is sufficient and revealing.

Professor Andrade confirms this underestimation of fieldwork by both quantitativists and Marxists,

> These essays seek to emphasise that theory and the search for a theme in Geography are deeply linked to fieldwork, which is so undervalued by quantitativists and Marxists, and that, when working with Geographical Science, we must always be concerned with a series of dualities, such as space and time, global and local vision, theoretical and empirical aspects, technology and direct observation, scientific competence and social commitment, etc. Science only exists as such when it is committed to society and seeks to fulfil social demands. The phase of the Ivory Tower and the dominance of personal interest over social interest is over. Nor can it be viewed in isolation, because there are not several sciences, but only one science that is fragmented in its approaches for methodological reasons. Totality is therefore essential to true scientific thinking, hence the need to give greater support to interdisciplinarity (1999, p. 18-19).

Corroborating with Chaveiro,

> There is no single way of doing geography, nor is there a single way of interpreting the main concepts that geographers work with. Not even the Geographical Renewal Movement was able to create a monolithic science based on a single theoretical and methodological parameter (1996, p. 44).

We understand that in Moraes' thinking, there is a dissonance in relation to the Traditional Geography, then,

> From 1970 onwards, Traditional Geography was definitively buried; its manifestations from that date onwards sounded like survivals, remnants of a bygone era. A time of criticism and proposals within this discipline is firmly established (1998, p. 93-94).

We see that the "gelatinous" solidity of Critical Geography defended by Moraes (1998, p. 93-94) ends up fading and reinforcing the thinking of Marshall Berman, in his 1998 work "EVERYTHING THAT IS SOLID FALLS INTO THE AIR".

Santos (1997, p. 9) says that "Critical Geography, which flourished during this period, cannot be satisfied with being merely critical. In order to be useful and utilised, criticism has to be analytical and not just discursive".

According to Silva (1996, p. 14), "today's Geography is a Geography made up of geographies that are related but do not form a unity".

In tune with Professor Andrade on Critical Geography,

> Some geographers, uncommitted to the society in which they live, have developed a destructive critique of everything that has been done and, in the name of a false materialism, present highly idealistic postulations that condemn the simple analysis of physical-natural factors and their importance in the formation and transformation of space, in the name of Marxist principles. A reading of the work of the founders of Marxism indicates that they were deeply familiar with the precepts of classical German geography, above all Humboldt, and although they sought to mitigate geographical determinism, they recognised the strong influence of the natural environment on relations between man and nature. Self-determined "critics" need to self-criticise and see that many of the criticisms they make of others should be made of their own work, they need to deepen their studies and reflections, consolidate their philosophical, ideological and political convictions and reflect on the world we live in, looking for solutions that can be achieved. Social commitment must be greater than personal interest, given that we are living in the last decade of the Second Millennium, one of the moments of society's most acute transformations, and Geography was born not only as a social science, but also as an eminently political science (1999, p. 55-56).

Finally, we understand that fieldwork is beneficial and fulfils its scientific and pedagogical objectives. Therefore, we say that it has been used for various purposes by peoples, scientific currents and geographical schools; and it is up to the geographer to intone the significance of this activity, where the field is a laboratory.

In this sense, in order to handle this open-air laboratory, the geographer has to look at and size up the theoretical, physical and human nuances learnt in the classroom or office very clearly.

We see the municipality, the place or the surroundings as the starting point for fieldwork to fulfil its objectives, articulate theoretical and practical interactions and establish reflections.

Having explained in chapter 1 how the fieldwork inside the

Geographical Science and its purposes in each historical context. In Chapter 2, we have to highlight the use and occupation of the Cerrado, as well as higher education in Goiás, the emergence of higher education units, municipalities and Geography courses in both Catalão and Pires do Rio.

CHAPTER 2

CATALÃO AND PIRES DO RIO: DIFFERENT CONTEXTS IN THE CERRADO OF GOIANO

The purpose of this chapter is simply to explore the Cerrado in a subtle and superficial way, to present a physiographic panorama, transformations, negative environmental impacts and occupations in this Biome, paying attention to the emergence of Catalão (mining period, 18th century) and Pires do Rio (railway period, 20th century). It also highlights higher education in Goiás and then historicises, in "quick brushstrokes", the scenarios of the implementation of higher education in Catalão (1980s) and Pires do Rio (1990s), highlighting the Geography Degree Courses in both cities.

2.1. Cerrado: physiography, use, occupation and degradation

In the work of Ab'Sáber (1995, p. 77-85; 2003, p. 40), the core area of the Central Chapadões Domain with Cerrados, Cerradões and Campestres with veins of Gallery Forests, is located predominantly on the Brazilian Central Plateau (figure 1); it is the second national biome, represents 22% of Brazil's territory with around two million square kilometres and is considered a water tank (cradle of water), as it stores and supplies the Paraná, Amazon, Araguaia/Tocantins and São Francisco river basins. This biogeographic formation is predominant throughout the Brazilian Centre-West and its surroundings, such as the south of Piauí and Maranhão, the interior of Tocantinense, the west of Bahia, the west and north of Minas Gerais, the north of São Paulo and enclaves in the Amazon and small areas of the north and north-east of Paraná (figure 2).

In order to characterise it biogeographically, the question arises: What is the Cerrado? In the work by biologists Franco and Uzunian (2004, p. 9), we find some notes:

> Cerrado, in Portuguese, has the meaning, among others, of "closed". In biological terms, however, it corresponds to an important Brazilian ecological formation. When you walk through the cerrado, in its most common form, you realise that it has low, crooked, thick-barked trees and that, on the ground, it contains a carpet of low plants, called herbaceous (from the Latin, herbaceu = herbs), which commonly belong to the grass family (the grass group, which botanists prefer to call the poaceae family). In other words, in technical terms, the cerrado usually has only two strata (layers, floors): the arboreal/shrubby, with a predominance of arborescent species, and the herbaceous/subshrubby, made up of grasses, other herbaceous species and small shrubs. Similar physiognomies can be found in the savannas of Africa.

Figura 1: CERRADO CORE AREA IN BRAZIL
Source: www.wwf.org.br, 2007
Organisation: CARNEIRO, V. A. (2008)

Figure 2: THE CERRADO IN THE BRAZILIAN SCENARIO
Source: EMBRAPA-Cerrados, 2007
Organisation: CARNEIRO, V. A. (2008)

Biogeographer Troppmair (1989, p. 97) considers the Cerrado biome to be a "trophic or savannah vegetation, always associated with a tropical climate with alternating dry and wet seasons".

In the Cerrado, the appearance of crookedness, thick bark and low trees is linked to the pedological system, which is highly acidic and nutrient-poor.

In this way, Lepsch's ideas (2002, p. 134-135) underpin what has been said above about the pedological system:

> Latosols under cerrado vegetation are acidic and poor in nutrients. This acidity (related to toxic aluminium) and lack of nutrients are some of the main causes of the appearance of the cerrado as natural vegetation, rather than forest. Despite the low natural fertility, most parts of the latosols in these areas can be used for intensive agriculture, as long as the harmful acidity is neutralised with the application of lime and adequate amounts of nutrients are added with the application of fertilisers.

The geographer Chaves (1998, p. 13), in his master's thesis, says that "the word Cerrado, which means 'closed', 'dense', can be transposed as a characteristic of homogeneous and uniform vegetation. Thus, the term cerrado became known generically as the characteristic vegetation of the Central Region of Brazil".

Continuing along the path of Chaves' ideas (1998, p. 13), as "the regions of Brazil, the popular nomenclatures for the natural physiognomy of the Cerrado, gain ground and certain denominations emerge such as tabuleiros, chapadas, campos, gerais and sertões".

Going back a little to the use of the term "water tank", we understand that it is due to the existence of several waterholes in sedimentary geological structures on fault lines, fractures and diacrasis that feed the hydrogeographic system in the Cerrado.

In this respect, Nascimento's work (2001, p. 13) emphasises that "the Cerrado Brazil is structured on the central plateaus of Brazil, with altitudes ranging from 500 to 1,600 metres above sea level, with myriad springs that form watercourses, distributing them to almost all of the country's major basins."

The Goiás NGO Forum for the Environment (1991, p. 11) points out that "Brazil's central plateaus, covered by the cerrado phytogeographic and morphoclimatic domain, constitute the summit of Brazil and also of South America, as they distribute a significant amount of water that feeds the continent's main hydrographic basins".

The Cerrado is made up of a mosaic of phytophysiognomies that vary from grassland formations, veredas to forest formations. The main factors that determine the type of vegetation cover that occurs in each location are: the availability of water, climatic conditions, the soil and the availability of nutrients.

The predominant climate is typical tropical, with a dry winter to rainy summer season.

According to Barbosa (1996, p. 11), the Cerrado Domain is:

> Located in the central highlands of Brazil, where tropical climates of a sub-humid character prevail, with two seasons: one dry, the other rainy. It constitutes the great Domain of the Humid Tropics, covered by a landscape that constitutes a mosaic of physiognomic types and ranges from grasslands to forested areas.

The pedology of the Cerrado is generally characterised by deep, bluish, dark red or yellowish red, porous, permeable, well-drained and intensely leached soils. We also find stony and shallow soils on slopes (lithosols and/or neosols), sandy soils (quartzarenic), organic soils (organosols), hydromorphic soils and ferruginous concretions (cangas and/or laterites).

The relief and geology of the Cerrado, located on the Central Plateau, is characterised by a crystalline core subjected to successive denudational cycles, with dissected relief, ancient Precambrian terrain, metamorphic lithology and topography flattened by erosive activity, which gave rise to the plateaus.

The Brazilian Cerrado began to be populated around eleven thousand years ago, with hunters and gatherers who adapted to the physical environment. Then came the indigenous people, who began to carry out a variety of agricultural activities until the 18th century, when the area was initially occupied with the opening and settlement of villages for mining activities (gold and precious stones) and extensive livestock farming, which led to the emergence of numerous towns and concomitant environmental degradation (PINTO, 1993, p. 11; NASCIMENTO; CASSETI, 1999, p. 16; PALACÍN; MORAES, 1994, p. 15-25).

According to Ferreira (2001, p. 25),

> The Cerrado has been occupied in a disorderly manner, at an accelerated pace that seems

> to go far beyond the capacity of its natural and artificial subsystems to resist and recover. The outlook for the Cerrado Biome therefore looks bleak, as nothing seems to escape the greatest threat from this perverse model of human-nature interaction.

It wasn't until the 20th century that we saw radical changes in the Cerrado landscape: the arrival of the railway, the implementation of the MARCH TO THE WEST programme launched by Getúlio Vargas (President of the Republic, 1930s) to occupy demographic voids, the struggle waged by Pedro Ludovico (Interventor/Governor, 1930s and 1940s) to found Goiânia, the struggle of Juscelino Kubitschek (President of the Republic, 1960s) to build Brasilia, the opening of roads, the appreciation of land, the emergence of extensive agriculture (monocultures) based on the liming technique (use of limestone), the expansion of the agricultural frontier, intensive mechanisation and urban intensification.

Nascimento and Casseti (1999, p. 16) show that:

> Two factors promoted the most recent agricultural expansion in the Cerrado: the construction of the new federal capital at the end of the 1950s and the adoption of development strategies and policies and investments in infrastructure between 1968 and 1980. The construction of Brasilia and a road system linking it to the country's dynamic core allowed the Cerrado to be opened up and occupied, resulting in the expansion of commercial agriculture from the 1970s onwards.

According to Pinto (1993, p. 11),

> Until the mid-1950s, the region remained practically isolated from the most populous and economically dynamic areas of Brazil. This isolation was mainly due to the lack of transport links. However, the establishment of Brasilia as the country's administrative centre in April 1960 led to radical changes in the Cerrado's landscape, with marked consequences for its physical, biological, social and cultural aspects. The old towns have rapidly become centres of development, where agriculture and services have a prominent place.

The arguments put forward by Nascimento and Casseti (1999) and Pinto (1993) focus on the construction of Brasilia as a magnet for investment and land occupation.

We think it's worth asking about the participation of the railway in the south-east of Goiás, Getúlio Vargas' March to the West and the construction of Goiânia in this context, as both represent important historical moments for the Cerrado.

According to Peixinho (2001, p. 13),

> At the end of the 1930s, the so-called "March to the West" helped boost the population dynamics of Central Brazil. Even though its results were not as expected, it was important for signalling Brazil's internalisation. Even with these interiorisation movements, until the 1940s the Central-West region was considered by the IBGE to be a natural region. This picture changed with the move of the federal capital to the Central Plateau in the early 1960s. With the construction of Brasilia, a wide network of connections was created between the Centre-West region and the Centre-South, integrating Central Brazil into the country's most dynamic space. Within this integration of the Centre-West is the process of occupation of the Cerrado areas.

In Chaul (2000, p. 115),

> The second major expansion of Goiás' frontiers can be felt in the rise of farming and cattle-raising: first through the self-transporting ox; second through the railway tracks. Both, each in their own time, brought new territorial features to Goiás, opening up roads, expanding spaces and shaping the regional economy.

Far removed, apparently, from the Cerrado environment, the coffee planted in the Paraíba Valley (the eastern area between the Mantiqueira and Mar mountains in São Paulo) provides economic support for the country and the railway soon expands to the corners of Goiás.

The penetration of the railway into the south-eastern area of Goiás led to vigorous population growth, economic development and the emergence of many towns, such as Pires do Rio, founded on 9 November 1922.

This geo-historical context, according to Chaul (2000, p. 120),

> The process of insertion into the national market only began with vigour around 1912. Two factors were of fundamental importance in explaining this phenomenon: the development of the coffee economy in the Centre-South of the country and the penetration of the railway tracks into Goiás territory.

The same author (2000, p. 124) explains that:

> Vargas' developmentalist march and its mirror in Goiás, Pedro Ludovico, therefore needed a capital that reflected the progress of the new Brazil, unveiled in 1930 and realised in 1937. A capital that would coordinate political life and stimulate economic life. A capital that would project Pedro Ludovico nationally and, in agreement with the federal government, enable the occupation of the Brazilian interior.

According to Dias (1992, p. 7), the Cerrado:

> Despite the soil and water restrictions, thanks to studies into soil management through liming, fertilisation and irrigation, and the good topography and texture, low cost of land, good road network and proximity to consumer centres, the cerrados have become the country's new agricultural frontier over the last two decades, to the point where the cerrado is now one of Brazil's largest grain producing regions and is recognised as the world's last great agricultural frontier.
>
> Unfortunately, this economic occupation of the cerrados has taken place without proper planning: the cerrados are seen by planners, financiers and farmers only as land to be occupied, i.e. the cerrado is only used as a substrate for agroforestry activities based on the planting and breeding of exotic species, as if there were nothing usable in this huge region. Native forests are cut down for firewood and charcoal, native pastures are eliminated and native animal populations are predatorily depleted. There is, of course, nothing wrong with using exotic species with proven and economical performance, but this should not be to the detriment of the heritage of our native fauna and flora already adapted to the peculiar conditions of the Cerrados.

Arbex Jr. and Olic (1996, p. 7-8) report that in the core area of the Cerrado:

> Its initial occupation dates back to the 17th century, in the early days of Portuguese colonisation, when the discovery of gold and precious stones attracted large numbers of Portuguese-Brazilians in search of these riches. With the decline of gold mining, the Midwest suffered a long process of economic and demographic stagnation. This situation changed significantly from the 1940s onwards.
>
> In this decade and the following one, the region suffered the effects of the expansion of São Paulo's agricultural activity, which went beyond the limits of the state and brought profound changes to the economic relations of the southern part of the Centre-West. The 1950s and 1960s were marked, firstly, by the process of building the country's new capital; then, by the spatial consequences of the establishment and development of Brasilia and the installation of the infrastructure created to put it in contact with the other areas of Brazil.
>
> In the following two decades, the biggest impacts on territorial and socio-economic structures were caused by the actions of the federal government, which, through regional and sectoral development plans and projects, made it possible to expand the existing infrastructure even further, stimulated direct and indirect migration, created exceptional conditions for the purchase of large tracts of land, encouraged and funded research into new planting and land use techniques, etc.

With the intensification of the modernisation of agriculture through the Revolution

From 1970 onwards, through government programmes (SUDECO - Superintendence for the

Development of the Midwest (1967), EMBRAPA/CPAC - Brazilian Agricultural Research Corporation/Centre for Agricultural Research in the Cerrados (1960), POLOCENTRO - Cerrado Development Programme (1975), and finally PRODECER/JICA (1979) - Programme for the Agricultural Development of the Cerrado Region/Japan International Cooperation Agency), the Cerrado is "battered" in a contumacious way for the sake of the world economy, underpinning increased productivity, the strengthening of monoculture, the concentration of land and income and socio-environmental marginalisation.

Geographers Mendonça and Thomaz Jr. (2004, p. 99) emphasise that:

> From the 1960s onwards, a process of change began in the use and occupation of land in the Centre-West, with the implementation of modern technical methods for growing grains and raising cattle. The traditional Cerrado areas - extensive plateaus with flat topography, which until then had been little used - began to be intensively exploited, due to the availability of capital (government programmes), technical resources (machinery), technology (development of scientific research) and support for the construction of infrastructure by the Brazilian state, as a way of enabling the interests of national and transnational private capital. These factors, combined with the state's credit and tax policies for the indiscriminate "rational occupation" of Cerrado areas and the construction of the necessary infrastructure, have made this region the country's agricultural "granary".

The Cerrado is characterised as a grain-producing area, mainly soya and maize, which fostered Brazil's agricultural modernisation process from the 1970s onwards.

From the 1980s onwards, we find cotton plantations on the plateaus of the state of Goiás, and today (since 2000) we are experiencing the invasion of the land by sugar cane plantations. It's an economic model geared towards the export of raw materials, supported and based on latifundia and monoculture activity.

The researchers Szmrecsányi et al. (2008, p. 99) point to five environmental impacts produced by the expansion of the current sugarcane monoculture:

> 1- General damage to the landscape and biodiversity;
> 2- Impacts on water and soil due to the intense use of agrochemicals, especially herbicides;
> 3- Impacts resulting from the excessive use of untreated vinasse;
>
> 4- Damage caused by fires, which precede the harvest;
> 5- High water consumption in the industrial processing of sugar cane.

One of the main consequences of this economic model (known as the modernisation of agriculture) adopted for the Cerrado environment is environmental degradation. In Chaves (1998, p. 33),

> The so-called "Green Revolution", developed internationally under the patronage of private foundations and the governments of rich countries, was a strategy that was conveyed ideologically as a systematic contribution to fighting hunger in the Third World. With a neo-Malthusian slant, the project aimed to incorporate areas of South America, Africa and Asia in order to produce grain.

In agreement with Medeiros (1998, p. 127), "the advent of the Revolution

The Green Revolution, which took place mainly in the 1970s, with the adoption of new technologies used to increase agricultural productivity, contributed substantially to the deterioration of the environment". Until the 1950s, the Cerrado didn't undergo profound changes in its landscape, and it was only from the 1960s onwards, with the construction of Brasilia and the implementation of the road plan, that it began to be exploited, giving way to the agropastoral activities of southerners and southeasterners.

According to Chaves (1998, p. 34),

> The culmination of the National Integration project came with the construction of Brasilia on the Brazilian Central Plateau, in the core area of the Cerrado. The transfer of the political/administrative axis from the Centre-South to the Centre-West of Brazil creates, from 1960 onwards, a dynamic axis determining a growing flow of capital to the Cerrado Region.

Pinto (1993, p. 12) emphasises that:

> From an ecological point of view, however, the disorganised way in which Cerrado lands have been effectively occupied is no different from that observed in other regions. This is because all the agricultural technology being adopted in the Cerrado responds to a model of agriculture geared towards immediate profit, with little or no concern for long-term conservation. Extensive continuous areas are being deforested to plant agricultural monocultures, without reserving samples of natural ecosystems that can act as a genetic bank and refuge for fauna and flora. Even footpaths and riparian forests are being severely damaged.

In this panorama of appropriation and occupation of the Cerrado, there has been an intensification of the agricultural frontier, deforestation for charcoal production, pasture and commercial crops, atmospheric pollution from fires, silting up, soil impoverishment, the burying of springs and footpaths, the disappearance of biodiversity, animal trafficking, the spread of erosion processes, the contamination of water tables, rivers and the Guarani Aquifer by pesticides, unbridled urban sprawl, the opening up of access roads, the construction of dams, the increase in rubbish production and the proliferation of

vectors harmful to man. The environmental situation described is corroborated by Dias (1992, p. 78), who states that:

> In addition to the expansion of the agricultural frontier, other phenomena have threatened the integrity of the Cerrados' ecosystems and renewable natural resources: the construction of large dams (e.g. the Grande, Paranaíba and Tocantins rivers) and roads, mining, pesticides and urban sprawl.

Ferreira (2001, p. 28-29) also emphasises that:

> We believe that the Cerrado can and should be explored, but this occupation must be based on studies aimed at a balanced coexistence between the geo-environmental and anthropogenic components. Above all, we highlight the accelerated degradation of water systems, especially the Veredas subsystems, which are being transformed into dams to supply projects to occupy and degrade the Cerrado Biome. We realise that very soon we will no longer have running water in our streams because we are "drowning our springs".

Environmental concerns about the Cerrado are diverse and Mesquita (2005, p. 2122) in his article warns:

> The Centre-West Region is home to the core areas of the Cerrado Biome, where the main springs of Brazil's great hydrographic basins are concentrated. Together with the veredas and the cerrado, these form the cradle of the South American continent's waters. But the cerrado's waters are threatened, both quantitatively and qualitatively, by anthropogenic

action, through uncontrolled deforestation, destruction and misuse of the veredas, clandestine irrigation with centre pivots, indiscriminate use of pesticides, direct discharge of chemical and industrial effluents and untreated urban sewage. And adding to these negative effects, the construction of dams for the purpose of generating electricity creates artificial environments, drastically altering the water, physical-chemical and biological quality, compromising the waters of the Cerrado.

Research carried out by EMBRAPA/Meio Ambiente (headquarters in Jaguariúna, state of São Paulo) in the Cerrado in the Guarani Aquifer domain by its researchers Gomes et al. (2006, p. 6) found that:

> This scenario, combined with the high natural vulnerability of the direct recharge areas of the aquifer in question, places them in a situation of high exposure to the risk of contamination, both of the water table and the deep water table, as well as favouring the formation of gullies and gullies, mainly due to inadequate agricultural practices. Given this scenario, agricultural research, particularly through Embrapa Meio Ambiente, has been carrying out work to assess the risks of contamination of the Guarani aquifer's groundwater, taking into account the different agricultural activities along its direct recharge or outcrop areas in Brazil. Other actions for these areas are also underway, including the so-called Good Agricultural Practices (GAPs).

The situation of the Cerrado Biome is extremely worrying, as researchers Alho and Martins (1995, p. 52) and Chaves (1998, p. 64) show that "the Cerrado has a great potential for wood, which has led to the destruction of native vegetation for the production of charcoal and the supply of steel industries in the state of Minas Gerais".

Many studies carried out in the Cerrado point to various negative impacts and Verdesio (1993, p. 589-591) highlights some of these environmental problems in his work: "genetic impoverishment, soil compaction and erosion, chemical contamination of water and biota (fauna and flora of a region) and the issue of irrigation".

With regard to soya cultivation, the environmental impacts presented by Medeiros (1998, p. 134-141) are: "loss of biodiversity, soil erosion and compaction, contamination of the environment by pesticides and social consequences (inability to create jobs in the countryside, rural exodus and wage labour)". Despite its valuable biogeographical importance, the occupational process in the Cerrado, especially in the last 50 years, has led to the unbridled destruction of this biome. Human occupation follows the same dictates and characteristics that have characterised economic history in other parts of the country and which have led to the devastation of areas of Atlantic Forest, Araucaria, Caatinga, Mangroves, Amazon Rainforest, among other vegetation groups.

We need to reverse this impactful geo-environmental situation in the Cerrado, through legislation, environmental education, sustainability and public policies, before it's too late and nothing can be done.

Medeiros (1998, p. 143) emphasises that:

> The search for perfect harmony between new production techniques, agricultural policies and agroecological conditions is the best and safest way to avoid wasting resources and production losses, reduce costs and promote sustainable coexistence between man and nature.

Within this physiographic framework of uses, occupations and degradations in the Cerrado, the municipalities of Catalão (Geographic Microregion-17) and Pires do Rio (Geographic Microregion-16), in the state of Goiás, have different historical trajectories, which we will now learn a little about (figure 3).

It is worth remembering that the municipalities of Catalão (item 2.2.1) and Pires do Rio (item 2.2.2) are located in south-eastern Goiás and respectively have higher education institutions: UFG-Universidade Federal de Goiás and UEG-Universidade Estadual de Goiás (map 1).

Next, we need to know a little more about higher education in Goiás (item 2.2) and how these colleges were established in the interior of the state, especially in Catalão and Pires do Rio.

1- São Miguel do Araguaia; 2- Rio Vermelho; 3- Aragarças; 4- Porangatu; 5- Chapada dos Veadeiros; 6- Ceres; 7- Anápolis; 8- Iporá; 9- Anicuns; 10- Goiânia; 11- Vão do Paranã; 12- Entorno de Brasília; 13- Sudoeste de Goiás; 14- Vale do Rio dos Bois; 15- Meia Ponte; **16- Pires do Rio**; **17- Catalão** e 18- Quirinópolis.

Figure 3: The Microregions of the State of Goiás Source: SEPLAN/SEPIN (ESTADO DE GOIÁS), 2007 Organisation: CARNEIRO, V. A. (2007)

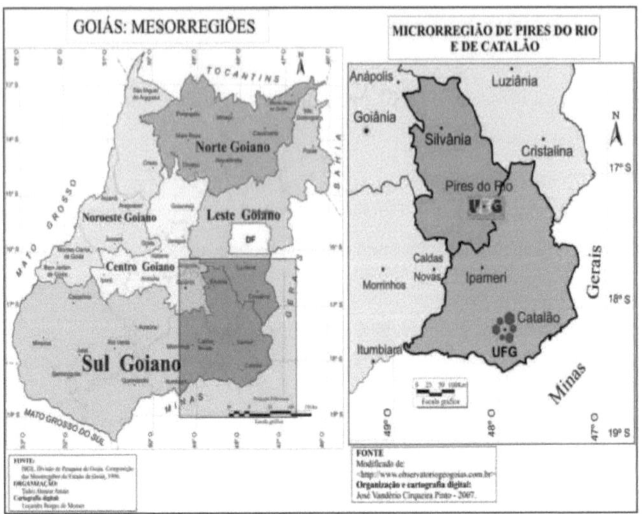

Map 1: Location of UEG - Pires do Rio and UFG - Catalão in south-eastern Goiás Sources: IBGE and Observatório de Goiás (2007)
Organisation: PINTO, J. V. C., 2007

2.2. Higher education in Goiás

In this first moment, before diving headlong into the issue of higher education in Goiás, we need to look at the historical background of education in Brazil.

Professor Melo states that the teaching of geography:

> Long before it was institutionalised as a school subject and as an academic science, geography was already strategic knowledge at the service of political leaders and businessmen. In Brazil, specifically, it was also used before the institutionalisation of schools, highlighting the importance of geographical knowledge in the lives of the first holders of knowledge on national soil, the Portuguese colonisers (2001, p. 25).

The starting point was the establishment of the religious education system by the Society of Jesus, which was founded by Ignatius of Loyola in 1534. From then on, the Ratio Studiorum (study plan) was implemented by the first Jesuits on Brazilian soil in 1549 and, concomitantly, in the other colonies controlled by Portugal (MELO, 2001, p. 25).

According to Ribeiro (1993, p. 26), "the Jesuits followed the guidelines contained in the

Ratio whatever the region in which they worked". In this way, the Jesuit priests are considered to have been primarily responsible for education in the Brazilian colony for more than two centuries (1549-1759) and also contributed to the process of colonising Brazil.

Another important detail of colonisation relates to the Metropolis-US Pact.

Colony that did not allow the establishment of educational, cultural and other forms of knowledge immediately, because it was necessary to maintain the submission of the natives to the new regime.

According to Niskier (1989, p. 41),

> The foundation of Vila de São Vicente (on the São Paulo coast) by Martim Afonso de Sousa on 22 January 1532 marks the effective start of Portuguese colonisation in Brazil. However, when this first permanent nucleus of settlers was created, nothing related to the education of its inhabitants was considered.

During the Jesuit period, the Jesuits preached the Catholic faith and introduced European customs to educational work via primary schools. This missionary task ran from 1549 to 1759 (MELO, 2001, p. 25-26, RIBEIRO, 1993, p. 22-23). In Piletti and Piletti,

> The Society of Jesus was founded by Ignatius of Loyola in 1534 as part of the Catholic Church's reaction to the Protestant Reformation. Its main objective was to halt the Protestant advance on two fronts: a) through the education of the new generations; and b) through missionary action, seeking to convert the peoples of the regions that were being colonised to the Catholic faith (2003, p. 134).

Ribeiro emphasises that "the intellectual education offered by the Jesuits, and therefore the education of the colonial elite, was marked by an intense 'rigidity' in the way of thinking and, consequently, of interpreting reality" (1993, p. 25). Continuing with the author,

> The adoption of the guidelines for the administration of material goods contained in the "Constitutions" is yet another indication of how this union between the Portuguese government and the Jesuits was conducted to the greater benefit of the latter. This subsequently led to a clash, culminating in the expulsion of the Society of Jesus from Portugal and Brazil in 1759 (1993, p. 27).

This shock "shows a lack of interest or the realisation that it was impossible to 'educate' the Indians as well" (RIBEIRO, 1993, p. 22). From this point of view, "it was necessary to concentrate personnel and resources on 'strategic points', since there were few of them. And these 'points' were the children of the settlers to the detriment of the Indians, the future priests to the detriment of the laity, justifying the religious" (RIBEIRO, 1993, p. 22). Thus, "the legal plan (to catechise and educate the Indians) and the real plan are far apart. The educated will be descendants of the colonisers. The Indians will only be catechised". (RIBEIRO, 1993, p. 23).

According to Niskier, the Marquis of Pombal, "the new minister, maintained good relations with the Jesuits after taking office in 1750, which only changed when work began on demarcating the borders in southern Brazil" (1989, p. 56).

The conflict in the southern region of Brazil was against the rebellious Guarani Indians, precisely in the area of the Seven Peoples of the Missions, which greatly delayed territorial demarcation. This was the trigger for the enactment of the law of 3 September 1759, which established the expulsion of the Jesuits from the colonies and from Portugal (Niskier, 1989, p. 5658). The period was marked by the expulsion of the Jesuits in 1759 by the Marquis of Pombal; the educational structure had 25 residences, 36 missions, 17 colleges and seminaries, as well as small seminaries and schools of first letters built in the main Brazilian cities where there were establishments of the Society of Jesus (NISKIER, 1989, p. 58).

With this, a new scenario was established in which "from then on, secondary education, which at the time of the Jesuits was organised in the form of a course - Humanities, was now organised in separate classes (royal classes) of Latin, Greek, philosophy and rhetoric"

(RIBEIRO, 1993, p. 34).

From a pedagogical point of view, the Pombaline teaching project was a step backwards and seen from another angle, it presented new methodologies and new books (RIBEIRO, 1993, p. 34). But, according to Romanelli, "with the expulsion, an entire administrative teaching structure was dismantled" (1986, p. 36).

Niskier (1989, p. 19) states that:

> There are those who attribute some questionable characteristics of our style to the educators of the Society of Jesus, who arrived here in 1549 with the arrival of the first Governor General, Tomé de Sousa: verbalism, academicism, bachelorism and other "isms". The lack of a more science-orientated education, which would give us better conditions for international competitiveness, is condemned. While this is true, it is still to the Jesuits' credit that our territorial integrity and the existence of a single, predominant religion were maintained, and that Portuguese remained the language used in the schools and churches set up along the Brazilian coast.

For three centuries, the Lusitanian metropolis did not take kindly to the professional or intellectual training of an elite in its colonies, including Brazil. According to Piletti and Piletti,

> History shows us that, despite the intense struggles of its people, Brazil has always been kept in a situation of dependence. Initially, from Portugal, then from England, and finally from the United States. And education was one of the instruments used by the successive groups that occupied power to promote and preserve this dependence. When not through outright exclusion, preventing most Brazilians from accessing school, through teaching for submission, devoid of critical concern, both in its content and in its methods (2003, p. 132).

It should be noted that the Jesuits who were in Brazil in the 16th, 17th and 18th centuries produced a lot of geographical material via fieldwork, which was taken to Europe. Freitas (2003, p. 31) states that:

> The letters, reports and works of the Jesuits, written in Brazil by those who lived here, brought to Europe important knowledge of the Brazilian land, through systematic empirical work that allows us to place them among the first geographers in our land and thus worthy of being placed in the light of geographical thought.

The same author (2003, p. 32-35) records three important moments in Jesuit works in Brazil:

> 1- The 16th century was the phase of the first "contact". During this period, the first "awe-inspiring" reports or descriptions of facts and things "unknown to this world" stand out. The vast majority of works from this period describe the climate, fauna, flora, nature and territory.
> 2- The Jesuit work produced in the 17th century was mainly concerned with indigenous ethnography and "reconnaissance" (geographical discoveries) in the territory (rivers, valleys, roads, etc).
> 3- This apprehension, "understanding", of Brazilian nature was materialised in the Jesuit works that still deal with the geographical discoveries of the 18th century or in the survey of coordinates and distances, as well as in the more and more sophisticated and improved cartographic drawing of Brazilian territory; or in the indigenous culture that was increasingly becoming an object of interest in Europe; in the animals and plants that were being known and classified and in many other themes developed in the Jesuit works. We can say that colonisation was coming to an end as we moved towards greater and greater assimilation of the territory, of nature, in short, of all the elements of this New World.

In Melo (2001, p. 26-27),

> Although Geography did not exist during the colonial period, it was present in a diffuse way in Portuguese texts, in which the characteristic geographical features of our country appeared. On the other hand, Geography was present in people's lives when, for example, the bandeirantes travelled into the "sertões", in addition to the constant updating of maps that the Portuguese Crown itself promoted.

The Pombal period, which lasted from 1760 to 1808, was marked by the implementation of the charter of 28 June 1759, a stage in the destruction of the Jesuit educational programme in Portugal and in all the colonial domains (MELO, 2001, p.26). For Romanelli, "the rise of the Marquis of Pombal, whose line of thought was closely linked to encyclopaedism, resulted in the expulsion of the Jesuits from Portugal and its domains" (1986, p. 36).

According to Piletti and Piletti (2003, p. 137), the Marquis of Pombal:

> He took several measures to centralise the administration of the colony in order to control it more efficiently: he abolished the system of Hereditary Captaincies, elevated Brazil to the category of a viceroyalty, transferred the capital from Salvador to Rio de Janeiro, etc.

The main idea of the Pombaline reforms was to replace the schools that served the Catholic Church with schools with strong ties to the state (PILETTI; PILETTI, 2003, p. 146). Ribeiro is quite categorical in stating that:

> Thus, it is clear that the "Pombaline Reforms" aimed to transform Portugal into a capitalist metropolis, like England had been for over a century. They also aimed to bring about some changes in Brazil, with the aim of adapting it, as a colony, to the new order sought in Portugal (1993, p. 35).

At the beginning of the 19th century, Brazilian education was "tottering" due to the collapse of the Jesuit educational programme, without any other kind of effective and coherent education system being organised in its place.

From 1808 to 1821, the Joanine period, when the Portuguese royal family arrived, there was a break with the previous situation. In order to meet the needs of his stay in Brazil, the monarch Dom João VI ordered the opening of military academies, law and medical schools, the Royal Library, the Botanical Gardens and the Royal Press (RIBEIRO, 1993, p. 40; ROMANELLI, 1986, p. 38; NISKIER, 1989, p. 82-92; MELO, 2001, p. 27).

We need to explain that:

> When Portugal was invaded (1807) by French troops and the royal family and court were forced to come to Brazil under English guard, the combination of these interests (colonial and English groups) forced the Prince Regent to decree the "Opening of the Ports" (1808), even though it was temporary, but in reality it was never revoked (RIBEIRO, 1993, p. 39).

For Piletti and Piletti (2003, p. 145),

> With the arrival of the Portuguese Royal Family in Brazil (1808) and Independence (1822), the government's main concern in terms of education became the training of the country's ruling elites. Instead of endeavouring to set up a national education system, integrated in all its levels and modalities, the authorities were more concerned with creating a few higher education schools and regulating the access routes to their courses, especially through secondary education and entrance exams to higher education studies.

In the imperial period, also known as the monarchy (1822 - 1888), King João VI returned to Portuguese territory, D. Pedro I proclaimed Brazil's independence and the first

Brazilian Constitution was granted in 1824 (NISKIER, 1989, p. 101). According to Ribeiro, Brazil's first constitution was in fact "inspired by the French Constitution of 1791 and was therefore much more radical in its proposals" (1993, p. 45).

Brazil's Magna Carta of 1824 established in its article 179 that primary education was free for all citizens, and that was it (RIBEIRO, 1993, p. 45; NISKIER, 1989, p. 101).

According to Piletti and Piletti (2003, p. 159),

> Free education already appeared in the 1824 Constitution. The 1891 Constitution said nothing about it, leaving it up to the states to take charge of primary education. Gratuity and compulsory education appeared together for the first time in the 1934 Constitution, which in article 150 established "free comprehensive primary education and compulsory attendance, extended to adults". Since then, the principle of free and compulsory education has never ceased to be present in our Constitution.

From the Empire until the Proclamation of the Republic in 1889, practically nothing was done coherently for Brazilian education, which, according to Piletti and Piletti, "the educational legacy that the Empire left to the Republic can be seen from two aspects: quantitative and qualitative" (2003, p. 151). The authors show that from the quantitative aspect

> At the end of the Empire, for a population of almost 14 million inhabitants, we had around 250,000 enrolled in primary schools. If we add to this those enrolled in all other courses, we arrive at close to 300,000 students, around 15 per cent of the school-age population (PILETTI; PILETTI, 2003, p. 152).

It was also emphasised that, in terms of quality, the Republic did not inherit the

The Empire had an articulated system of education: to enter secondary school, completion of primary school was not required; to enter higher education, completion of secondary school was not required (PILETTI; PILETTI, 2003, p. 152-153).

A landmark event in 1837 was the creation of Colégio Pedro II, which made many of its subjects compulsory, which led to Geography gaining the "status" of a subject in the Official Programme. In Niskier,

> Having been relieved of the burden of providing primary education in the provinces, the regency was able to approve, by Decree No. 14 of 15 March 1836, the regulation of primary schools in the Court and in the Neutral Municipality.
> Undoubtedly, the most important event in the field of public education in Brazil in 1837 was the creation of Colégio Pedro II, by decree of 2 December, an initiative of Bernardo Pereira Vasconcelos, during the Regency of Pedro de Araújo Lima, the future Marquis of Olinda (1989, p. 111).

He goes on to point out that the first three articles of the Decree read as follows:

> 1- St Joachim's Seminary is converted into a secondary school.
> 2- This school is called Colégio Pedro II.
> 3- This college will teach the Latin, Greek, French and English languages, rhetoric, and the elementary principles of Geography, History, Philosophy, Zoology, Mineralogy, Botany, Chemistry, Physics, Arithmetic, Algebra, Geometry and Astronomy (NISKIER, 1989, 111-112).

Melo's thoughts on this event are:

> This is yet another milestone in the history of Geography teaching in Brazil, which until then had been diluted in literary texts. Geography's "status" as a subject in secondary education was a real gain for the country. However, the content taught was not yet that proposed by Modern Geography, which was already being practised in Europe at the same time (Karl Ritter's Geography). Geography teaching in this period of Brazilian history generally consisted of a list of names and facts, i.e. nomenclature, easily found

> in the atlases and textbooks of the time (2001, p. 28).

The republican period (1889 - 1961) saw a strong presence of the positivist current in school organisation, which originated in Europe. According to Dourado,

> The policy of creating isolated courses implemented in the country from 1808, with the installation of the first higher education courses in Rio de Janeiro, suffered significant French ideological and cultural influence of a positivist orientation, the legacy of which certainly contributed to the delay in creating universities (2001, p. 14).

According to Melo (2001, p. 33-35), "Brazil went through a phase of major social, economic and political transformations, requiring pedagogical renovations", which left the teaching of Geography, as well as other subjects, at the mercy of the various educational reforms (Benjamim Constant, Epitácio Pessoa, Rivadávia da Cunha Corrêa, Carlos Maximiliano, Luiz Alves Rocha Vaz, Francisco Campos and Gustavo Capanema) that took place from 1880 to 1961.

Before discussing the post-1930 period that confirmed the institutionalisation of the Geography Chair in Brazil, it is worth highlighting the masterful contribution of Carlos Miguel Delgado de Carvalho as a Geography teacher at Colégio Pedro II, a promoter of fieldwork as a teaching resource and one of the creators of the Free Higher Course in Geography at the Geographical Society of Rio de Janeiro (MELO, 2001, p. 35-38; ANDRADE, 1999, p. 11; MONTEIRO, 2002, p. 6-7).

According to Sodré (1973 apud Ribeiro, 1993, p. 96), the phase prior to the Revolution of 1930 was known as the "decline of the oligarchies", because this decline evidently occurred due to the existence of new social forces as a result of changes in the economic structure.

Romanelli explains:

> Practically most of the social organism participating in or aware of the political and economic process that was taking place was unhappy with the system in place. The elites who remained in power were seeing their bases of support undermined day by day and the government was visibly losing its authority. The 1930 elections for the Presidency of the Republic, held in a climate of great political unrest and in which, as usual, fraud prevailed, gave victory to the candidate of the situation. This fact, combined with the assassination of the opposition candidate for the Vice Presidency, ended up being the concrete reasons needed for the armed movement to break out (1986, p. 49).

He goes on to say that

> The 1930 Revolution, the result of a crisis that had been destroying the monopoly of power by the old oligarchies for a long time, favoured the creation of some basic conditions for the definitive establishment of industrial capitalism in Brazil, and therefore ended up also creating the conditions for a change in the cultural horizon and the level of aspirations of part of the Brazilian population, especially in the areas affected by industrialisation. It was then that the social demand for education grew and became an increasingly strong pressure for the expansion of education (ROMANELLI, 1986, p. 60).

Historically, 1932 is a landmark year because a group of educators published the Manifesto of the Pioneers of New Education to the Brazilian population, written by Fernando de Azevedo and signed by renowned educators of the period, clearly stating that education is a right for all (ROMANELLI, 1986, p. 179; RIBEIRO, 1993, p. 106-117; PILETTI; PILETTI, 2003, p. 176-181; NISKIER, 1989, p. 253-256).

For Piletti and Piletti (2003, p. 177-178), the educators launched the Manifesto of the

Pioneers of New Education, pointing out:

> 1 - Education is seen as an essential instrument for the reconstruction of democracy in Brazil, with the integration of all social groups; 2 - Education must be essentially public, compulsory, free, lay and without any segregation of colour, sex or type of studies, and developed in close connection with communities; 3 - Education must be "one", with the various levels articulated to meet the various stages of human growth. But unity does not mean uniformity; rather, it presupposes multiplicity. Therefore, although the school is one and the same on the bases and principles established by the Federal Government, it must adapt to regional characteristics; 4 - Education must be functional and active and curricula must adapt to the natural interests of the students, who are the backbone of the school and the centre of gravity of the education problem; and 5 - All teachers, even primary school teachers, must have university training.

The authors state that "higher education underwent important changes from 1930 onwards. With the creation of the first universities, the school phase was overcome.

These were isolated higher education institutions with a markedly professional character" (PILETTI; PILETTI, 2003, p. 180). After the Constitutionalist Revolution of 1932, in 1934, the University of São Paulo (USP) was founded, centred on the Faculty of Philosophy, Sciences and Letters. This was the cradle of the emergence of geographical science in Brazil and the strong influence of the French school of Paul Vidal de la Blache and the pioneering work of Pierre Monbeig and Pierre Deffontaines in São Paulo.

At this point, it's worth noting that, according to Valverde (1994, p. 118), "what mattered to Pierre Deffontaines was bringing together those interested in geography and working with them, going on excursions and debating". As Pierre Deffontaines was the founder of the AGB - Association of Brazilian Geographers (1934), the researcher Andrade points out that "this period was very useful because it allowed the development of fieldwork and knowledge of various areas of the country, through research carried out at the AGB's general meetings" (1999, p. 11).

Returning to the subject of higher education in São Paulo, Professor Niskier (1989, p. 256) states that:

> Once the political situation in the state had stabilised after the serious problems generated by the Constitutionalist Revolution of 1932, and a "São Paulo civilian" interventor had been appointed to administer São Paulo, some of the most important popular aspirations could finally be realised. These included the creation of a university that would bring together the higher education establishments in the state capital into a single institution. On 25 January 1934 - the day that marked the 380th anniversary of the city's founding - interventor Armando Sales de Oliveira signed Decree No. 6.283, which created the University of São Paulo, based on the Faculty of Philosophy, Sciences and Letters.

Thus, in São Paulo, in 1934, on the initiative of Governor Armando Salles Oliveira, USP was created, following the pillars of the Statute of Brazilian Universities, Decree no. 19.851, of 14 April 1931 (PILETTI; PILETTI, 2003, p. 180; MONTEIRO, 2002, p. 10). And in 1935, Professor Anísio Teixeira created the University of the Federal District (now the Federal University of Rio de Janeiro) in Rio de Janeiro (RIBEIRO, 1993, p.176; MONTEIRO, 2002, p.

10).

The Statute of Brazilian Universities came into force with few adjustments and changes until 1968, when the university reform programme was established. In this Statute, Brazilian universities could be created and maintained by the Union, the States or, in the form of foundations or associations, by private individuals, constituting federal, state and free universities.

Niskier (1989, p. 316) states that:

> In 1930, there were only two universities in Brazil - a federal university in Rio de Janeiro and a state university in Belo Horizonte - to which reference has already been made. Other institutions of the same kind, such as the one in São Paulo in 1934, the one in the Federal District (the first) in 1935 and the one in Porto Alegre in 1936, were all created by local governments.
> However, the massification of higher education took place between the end of 1945 and the vote on the Law of Guidelines and Bases in 1961.

Conjecturing on the government of Juscelino Kubitschek, we have the following information:

> From 1960 onwards, Brazilian higher education underwent an unprecedented expansion. It all started with the changes that took place after 1950 and especially during the Juscelino Kubitschek administration. The labour market expanded to such an extent that demand, especially at university level, also kept pace with the president's developmentalist model.
> The rapid growth of companies and the state bureaucracy's need for highly educated staff, combined with the general desire to reach a higher position in society, made the expansion of higher education an imperative. The tertiary sector greatly absorbed the fruits of this expansion. Another factor was the parallel growth of the secondary school network, which began to put pressure on the third level with a demand for candidates that was always greater than the number of places offered by the entrance exams. The high point of this pressure came in 1968, when 125,000 candidates were refused places.
> As a result of this expansion, higher education ended up being mostly privatised. While in 1960, private institutions offered only 43 per cent of places, in the 1970s they offered 77 per cent (NISKIER, 1989, p. 383-384).

For this reason,

> The growing expansion of higher education after 1964 was effected by an increase in the private sector and in the number of places at federal HEIs and, in particular, by the expansion of public and private foundations, which gave this policy a clear character of massification of education as opposed to its democratisation (DOURADO, 2001, p. 15).

From 1945 to 1961, several universities were established in Brazil, including the Catholic University of Goiás in 1959 and the Federal University of Goiás in 1960 (NISKIER, 1989, p. 316). Cassimiro (1974, p. 200-202) makes it clear that:

> Higher education in Goiás is mainly represented by two universities: the Catholic University of Goiás, founded in 1959, and the Federal University of Goiás, founded in 1960.
> The Catholic university is made up of six teaching units that teach 17 undergraduate programmes: Geography, History, Law, Architecture, Arts, Economics, Business Administration, Accounting, Pedagogy, Maths, Physics, Natural History, Vernacular Letters, English, French, Social Work and Nursing. The Federal University is structured into basic knowledge units and applied knowledge units. The former are five in number and teach the following undergraduate courses in addition to basic studies aimed at candidates for the vocational area: Maths, Physics, Chemistry, Geography, History, Social Sciences, Vernacular Letters, English, French, Arts, Biology and Journalism. There are eight applied or professionalising knowledge units, which offer the following degree courses: Law, Dentistry, Pharmacy, Biochemistry, Engineering (Civil and Electricity), Medicine, Agronomy, Veterinary Science, Education (Pedagogy and Graduation) and an institute specialising in tropical pathology.

In the Centre-West of Brazil, today we have federal, state and free (private) higher education institutions. In the case of Goiás, we highlight the pioneering UFG - Federal University of Goiás (federal system, in 1960), UCG - State University of Goiás (religious system, in 1959) and a few private colleges. In the period before the UCG/UFG institutions, the 1960s, Dourado (2001, p. 43) and Baldino (1991, p. 55-56) point to the existence of a few colleges in Goiás, such as the Goyaz School of Law (1903), the School of Pharmacy (1922) and the School of Dentistry (1923), which came from the State Executive.

The first decades of the 20th century saw countless transformations in Goiás, as well as new ideas and technologies. In addition to the political changes resulting from the so-called 1930 Revolution, the arrival of electricity would change life in the cities forever, the railway would provide links with the rest of the country and the March to the West was officially established as a project to occupy the lands of the Cerrado, of which Goiânia would be one of the main landmarks. In Cassimiro (1974, p. 84-85),

> Despite the progress that the railway brought to the cities in the south of the state, it can be said that the First Republic did very little for Goiás, whose real phase of expansion and prosperity began in the 1930s.
> The founding of Goiânia marked an epoch in the political and economic history of Goiás. In 1937, a decree was signed to move the state capital to another location. With extraordinary sacrifices, few resources and a small loan of six contos de réis from the Bank of Brazil, the government of Pedro Ludovico Teixeira began building the new state capital. The main public buildings were completed that year, and the city of Goiânia was officially inaugurated on 15 July 1942, in a ceremony they called the "Cultural Baptism of Goiânia".

During this period, the Catholic Church promoted permanent and uninterrupted action in the backlands of Goiás in favour of the creation of educational institutions in the region. Thus, Dom Fernando Gomes dos Santos (Archbishop of Goiânia in the 1930s), requested studies for the establishment of a university in Central Brazil, which resulted in the emergence of the University of Goiás in 1959, and later, in 1972, it was renamed the Catholic University of Goiás (BALDINO, 1991, p. 66-79). This was followed in 1960 by the dream of the Federal University of Goiás, an arduous struggle by the Freemasons in Goiás (BALDINO, 1991, p. 66-79).

Researchers such as Cassimiro (1974, p. 202), Baldino (1991, p. 80-89) and Dourado (2001, p. 51) state that in addition to the two universities (UCG and UFG), there were four other isolated higher education institutions in the state of Goiás. 51) state that in addition to the two universities (UCG and UFG), there were four other isolated higher education institutions in the state of Goiás: a State School of Physical Education (1962) in Goiânia, a State Faculty of Economic Sciences (1961) in Anápolis and two more private institutions in Anápolis: the Bernardo Sayão Faculty of Philosophy (1961) and the Faculty of Law (1969).

Dourado (2001, p. 66), says:

> As far as higher education is concerned, this system expanded significantly in the 1980s in Goiás. Until 1979, the state had only two universities, UFG and UCG, and nine isolated higher education establishments (five private, three state and one municipal).

Baldino (1991, p. 119-121) called this period in the 1980s the "euphoric expansion of higher education in Goiás". In the 1980s, the Federal University of Goiás (UFG) launched a policy of internalisation through the establishment of Advanced Campuses in the municipalities of Catalão and Jataí (BALDINO, 1991, p. 119-121; DOURADO, 2001, p. 158). In the case of the Catalão Campus, it began operating with Geography and Literature programmes in 1986 (DOURADO, 2001, p. 168).

Long before the name UNIANA - Universidade Estadual de Anápolis (State Decree No. 3.549 of 1990) was established, Anápolis already had an Accounting Sciences course at the Faculdade de Ciências Econômicas de Anápolis (FACEA), which was created in 1961 by State Law No. 3.430 and granted its operation by Federal Decree No. 73.149 of 1973 (BALDINO, 1991, p. 80-89; UEG, 2008, not paginated).

Later, in 1984, FACEA offered courses in Administration and Accounting. And in 1986, the courses in Data Processing Technology and Short Degrees in Literature, History and Geography appeared (BALDINO, 1991, p. 80-89, UEG, 2008, not paginated).

UEG is the result of the transformation process of the State University of Anápolis - UNIANA and the isolated state colleges, including the Pires do Rio Teaching Unit, maintained by the State Government, via State Law No. 13.456 of 1999, which organically linked UEG to the State Department of Education, and later, through Decree No. 5.158 of 29/12/1999, to SECTEC - the Goiás Department of Science and Technology (UEG, 2008, not paginated).

Now, in fact, we're going to learn a little about the historical and geographical events in Catalão that contributed to the creation of the Geography Course and the Advanced Campus of the Federal University of Goiás in this town.

2.2.1. Getting to know Catalão

The municipality of Catalão (map 1) is located in south-eastern Goiás and is part of the Catalão micro-region (figure 4). According to Palacín and Moraes (1994, p. 59), its history is linked to the penetration of São Paulo bandeirantes at the beginning of the 18th century, who wanted to discover mineral wealth and capture indigenous people in the lands beyond the Paranaíba, Tocantins and Araguaia rivers.

In his work, Ramos (1984, p. 23) emphasises that:

> It is also true that the bandeirantes were orientated by the elevation of the hills, among which our Morro de São João stood out, and when they passed through here they stopped at the place known to this day as Córrego do Almoço, near the Ribeirão Pirapitinga, where "Catalão" built his "Sítio".

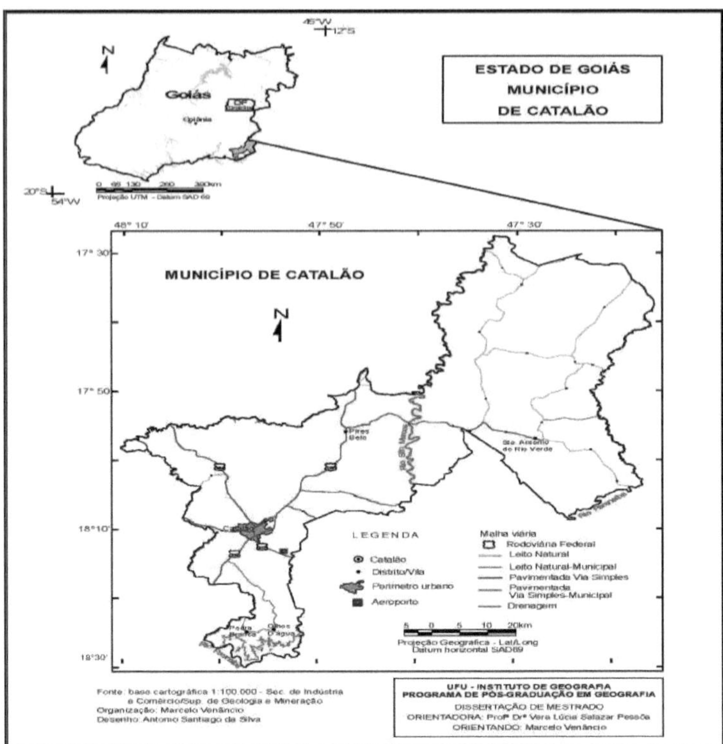

Map 2: Location of the Municipality of Catalão Source: VENÂNCIO, M. (2008)

Around 1722, Bartolomeu Bueno Filho's expedition passed through these lands and landed on the outskirts of Córrego do Almoço. In the meantime, a member of the entourage, a Spanish priest from Catalonia, nicknamed Catalão by everyone, deserted and put down roots at this stopping point (PALACÍN, 1994, p. 20). According to Campos (1979, p. 36),

> It is known that the Catalão landing harboured all the flags that passed through this region on their way to the flourishing gold mines of Goiás, in the regions of Pilar, Vila Boa de Goiás and the former town of Bonfim, now Silvânia.

Father Catalão, after whom the settlement was named, built a ranch by the Córrego do Almoço, one of the tributaries of the Ribeirão Pirapitinga, and from then on the area began to be settled in the middle of 1728. According to Campos (1979, p. 32):

> The entourage also included the manly figure of a Spaniard from Catalonia, nicknamed "The Catalan", who soon after crossing the Paranaíba River abandoned the flag. He decided to stay in this region near the Córrego do Almoço, marvelling at the privileged uber-friendliness of the soil and the excellent climate, and set up a landing point.

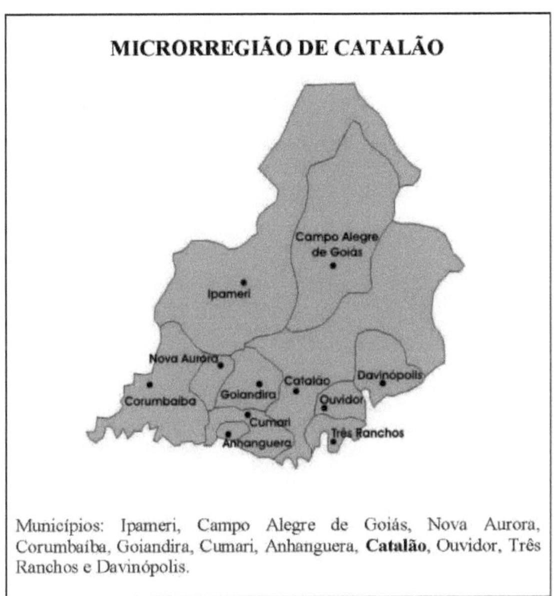

Figure 4: Catalão micro-region
Source: SEPLAN/SEPIN (STATE OF GOIÁS), 2007
Organisation: CARNEIRO, V. A. (2007)

Ramos (1984, p. 21) states that:

> Tradition, however, assures us that the founder was a simple companion of Bartolomeu Bueno Filho's "Bandeira"; a Spaniard or Spanish descendant of Catalan origin and nicknamed Catalão, the true founder of our secular community, who, after crossing the Paranaíba River, left the "Bandeira" and stayed here with other companions, including Friar Antônio, with the aim of building an inn, a point of support and reference for future explorers travelling between São Paulo and Goiás.

Bartolomeu Bueno da Silva (the son) left historical landmarks wherever he went and Catalão was no different, leaving a wooden cross (Cruz do Anhanguera) in the region known as Borda da Mata (Fazenda dos Casados) which, on 23/06/1916, was transported to the city of Goiás by authorisation of the Judge of Law of the Rio Paranaíba District, Dr Luiz Ramos de Oliveira Couto (CAMPOS, 1979, p. 32; RAMOS, 1984, p. 23-25).

On 20 August 1859, Francisco Januário da Gama Cerqueira, President of the Província de Goyaz, by Resolution no. 7, makes known to all the inhabitants of the Vila de Catalão, belonging to the Comarca do Rio Paranaíba, is elevated to the category of city, maintaining the same denomination (CAMPOS, 1979, p. 40-41; RAMOS, 1984, p. 23; PALACÍN, 1994, p. 32).

A few years later, in 1909, with the arrival of the railway tracks in Goiás, Catalão benefited from an increase in population and an expansion of economic activities. The railway gave a new lease of life to the Estrada de Ferro region, and Catalão was no exception, as Inocêncio (2006, p. 125) states:

> The establishment of a transport infrastructure in the south-east of Goiás was one of the first factors behind the main agrarian transformations that took place in the region from the beginning of the 20th century.
> The railway, which arrived in the south of Goiás around 1912, definitively and modernly integrated the state with the south-eastern region of the country. From this moment on, Goiás experienced one of its most prosperous phases in terms of agricultural development. There was no change in the type of product grown, but rather an expansion of the production area, where in some municipalities crossed by the railway, production went from subsistence to the most modern capitalist standards of cultivation in the whole country.

Chaul (1997, p. 158) explains the benefits of the railway system for Catalão:

> The city of Catalão, a major expression of the commercial and productive development of the south of Goiás in the First Republic, exemplifies the significance of the railways for the development and progress of this region when it is observed that the trade in agricultural production gradually grew.

Corroborating what Chaul (1997, p. 159) says, Catalão has become a reference point and "the demographic and economic rise of southern Goiás has transformed the region into one of the world's largest cities.

an example, for the time, of the state's progress and modernity, through its commercial and urban development". The same author (1994, p. 121) says that "in this way, the railway was the missing link in Catalão's commercial chain with the Triângulo Mineiro and São Paulo, for greater ease and growth".

According to Borges (1990, p. 87),

> Regional and inter-regional trade relations increased, developing an active import and export trade, with the emergence of significant commercial centres in the south-eastern region of the state that soon replaced the cities of the Triângulo Mineiro in controlling regional trade.

Thus, in addition to the railway issue, other geographical and historical factors, such as the founding of Goiânia (1930s), the emergence of Brasília and the opening of federal highways (1960s), the expansion of the agricultural frontier in the Cerrado (1970s) and the proximity of south-eastern Goiás to the south-eastern region of the country, all contributed to Catalão's rise on the Goiás scene. According to Venâncio (2008, p. 29):

> The seat of the municipality is located on the dynamic axis, which facilitates economic relations with other parts of Brazil. It has a wide range of transport facilities, including the BR-050 highway, which provides access to Brazil's major commercial centres, such as Brasília (DF), Uberlândia (MG) and São Paulo (SP), state highways and the Centro-Atlântica Railway (FCA), the former Federal Railway Network, now used exclusively to transport mineral and fertiliser production to the new agricultural frontier (Bahia, Mato Grosso and Goiás), via a railway terminal linking Catalão (GO), Araguari (MG) and Uberlândia (MG).

These characteristics created a favourable environment which, according to Venâncio (2008, p. 32), later contributed to "the establishment of mining industries in the 1970s and the modernisation of agriculture from the 1980s onwards, which placed Catalão on the country's trade balance". According to Gomes and Teixeira Neto (1993, p. 84), "in the south-east of Goiás, known as the Railway Region, the railway and the rare ores of Catalão-Ouvidor fostered a process of industrialisation that broke with the traditional and archaic economy".

Today, in the first decade of the 21st century, Catalão (photos 1, 2 and 3) has an economy based on activities in the automotive, agro-industrial, agricultural and commercial sectors and the mining of phosphate, niobium, titanium, vermiculite and rare earths.

This historical and geographical panorama led me to focus on the establishment of a higher education centre in Catalão (photo 4) via the inland expansion of the Federal University of Goiás (1980s), according to Silva (2005, not paginated), and the Geography Degree Course at this higher education institution, with its teachers and students, will be the target of my research.

Photo 1: Panoramic view of Catalão Author: CARNEIRO, V. A. (2008)

Photo 2: Partial view of Catalão
Source: MUNICIPALITY OF CATALÃO, 2007

Photo 3: Panoramic view of Catalão
Source: MUNICIPALITY OF CATALÃO, 2007

Photo 4: Partial view of the Federal University of Goiás in Catalão Source: SITE DO CAMPUS AVANÇADO DE CATALÃO/UFG, 2007

The following topic will discuss the struggle, a little behind the scenes, the Internalisation Project and the achievement of the Advanced Campus in Catalão with the implementation of the Languages and Geography Courses.

2.2.1.1, The creation of the CAC - Catalão Advanced Campus

The website of the Federal University of Goiás (2008, not paginated), Silva (2005, not paginated), Baldino (1991, pp. 80-89), Dourado (2001, p. 156) and Bretas (1991, pp. 596-597) state that the Federal Higher Education Institution in Goiás "was created on 14 December 1960 with the merger of five higher education institutions that existed in Goiânia (State of Goiás): the Faculty of Law, the Faculty of Pharmacy and Dentistry, the School of Engineering, the Conservatory of Music and the Faculty of Medicine". The same website (2008, not paginated) also states:

> Since then, Goiás has been able to train its own professionals and not depend on skilled labour from other regions of the country. For the young people of Goiás, this meant the opportunity for professional and intellectual training in a public, free and quality institution. It was a milestone in the state's history.
> In this vision, "the institution should be a centre for pedagogical, cultural, social and political transformation, inspired by culture and without a pre-conceived ideological conception," according to the then Rector Colemar Natal e Silva (at the time). The materialisation of this idea was the intensification of the University's cultural life and greater integration between students, teachers and the community.

From then on, the university gained prestige, space and strength and launched its "tentacles" towards the interior of the state of Goiás (Catalão, Firminópolis, Porto Nacional - which now belongs to the state of Tocantins, Cidade de Goiás, Rialma and Jataí) via the educational policy proposed by the MEC for public higher education institutions (DOURADO, 2001, p. 156).

The CAC-UFG Board Report (2008, not paginated) states that:

> In the incessant quest to internalise education in the state of Goiás, providing the poorest sections of the population with access to university, it is necessary for a public university to move out of the capital and into the countryside. With this in mind, UFG has agreements with the municipalities of Jataí, Catalão, Cidade de Goiás and Rialma to operate advanced campuses in these locations.

Silva (2005, not paginated) states that:

> The situation in the last years of the 1970s showed signs that the use of the repressive forces adopted by the military regime had been exhausted. The democratic transition was irreversible, as the subjective imbrication between national security and development gave way to the juxtaposition between democracy and development. The 1980s were marked by incessant debates between various social segments - neighbourhood associations, trade unions, student and teacher movements (unions and/or associations), all of which had in common the effort to guarantee the democratisation of society, public policies and social institutions.

According to Dourado (2001, p. 17), the 1980s were a fertile period for policies to expand and internalise higher education in Goiás, given that:

> The process of political redefinition in Goiás in the 1980s was marked by educational policies aimed at expanding higher education. During this period, there was an emphasis on the discourse of regional development and the need for expansion, understood in this context as the establishment of policies to internalise education. The creation of state colleges increased, especially in cities considered to be economic centres.

The same researcher (2001, p. 157) analyses the situation and makes the following points:

> The 1980s, the period in which UFG set up its advanced campuses, was a difficult time for the university, due to a reduction in funding, a ban on hiring professors, in short, a series of political, economic and institutional factors that made it difficult to operate.

According to Silva (2005, not paginated),

> Although the national scenario was not favourable to the expansion of higher education, given that public policies maintained a financial agenda for education that was somewhat derisory in relation to the demands in the area, the movement by UFG and the municipalities selected for the installation of advanced campuses was effective.

On this occasion, in 1981, the UFG Internalisation Programme was regulated by Resolution 156/1981 of the Teaching and Research Coordinating Council, with the aim of decentralising educational opportunities in higher education for young people from the interior of the state of Goiás (SILVA, 2005, not paginated; CAC-UFG DIRECTOR'S REPORT, 2008, not paginated, BALDINO, 1991, p. 118-119, DOURADO, 2001, p. 155-160).

The website of the Advanced Campus of Catalão/UFG (2007, not paginated) and the Report

of the CAC-UFG Board of Directors (2008, not paginated), show that:

> The Academic Unit of Catalão, which is part of the Federal University of Goiás, began as an Advanced Campus on 07/12/1983, by Decree No. 189. Its initial objective was to enable UFG to participate effectively in the process of local, regional and national cultural and socio-economic development. It was also intended to provide physical, administrative and technical bases for university extension programmes, linking the activities to be offered to the basic needs of the south-eastern region of the state of Goiás. In addition, the aim of the CAC/UFG was to provide qualified human resources in the various forms of extension work, with participatory action with local public bodies, offering qualification opportunities to teachers and workers from private companies at regional and local level, with the aim of attracting development to the region by improving qualifications in Catalão, which would very soon become a hub for agricultural, industrial and specialised services development.

According to Silva (2005, not paginated):

> UFG's Pro-Rectorate for Extension, through the Internalisation Programme, was promoting a policy of decentralising the university with the aim of expanding the academic functions of the Advanced Campuses, given that the Jataí and Catalão Campuses would be offering degree courses and with the prospect of offering courses in other areas. It is true that internalisation in the 1980s tended to fulfil a social function that was less tied to the ideals of the military regime, since the ideals of re-democratisation had been guiding social movements since the late 1970s, as well as the social practices of higher education institutions and other educational bodies.
>
> The internalisation of courses proposed by the university is based on the discourse of development and the democratisation of educational opportunities. From this perspective, the courses offered in the hub cities would focus on professionalisation, as they would be courses to meet the demands of the labour market and, in this case, UFG proposes to offer courses, but on a temporary basis, i.e. they would operate within the timeframe needed to train professionals according to the economic development demands of the municipality and region.

The same author (2005, not paginated) clearly states that:

> The 1960s were a watershed year for Goiás on the educational scene, with the creation of UCG (1959) and UFG (1960), complemented by the creation of the University of Brasilia in 1961. The creation of three universities in the Midwest in three consecutive years was no accident, but the result of a policy of internalisation initiated by Getúlio Vargas and concluded by Juscelino Kubitschek.

At the time, the main pillar of the Internalisation Programme was to emphasise the recommendations of the Ministry of Education and Culture, strengthening the improvement and implementation of extension courses and other types of higher education activities for teachers of primary and secondary education (formerly 1st and 2nd grade education), thus guaranteeing the pedagogical applicability of the results in the appropriate local communities (BALDINO, 1991, p. 210-211, DOURADO, 2001, p. 155-160). Silva (2005, not paginated) reports that:

> According to the UFG/1980 Interiorisation Programme, the MEC had recommended that our country's higher education institutions promote 3rd degree courses in qualification, training or retraining of staff who were working in education in order to improve 1st and

> 2nd degree teaching. In this sense, this UFG/1980 Interiorisation Programme had the scope to meet the MEC's recommendations, but in relation to the creation of the CAC, we still have two vectors that drove its creation: the first is that there was political will within UFG, since the Pro-Rectory of Extension at the time was in Catalão and Professor Arédio Teixeira Duarte, then a CONSUNI Councillor, is also a son of the land, and also the fact that there was political will in Catalão.

The Federal University of Goiás, through its Internalisation Programme (1980), sought to adhere to the MEC's recommendations and in the meantime the proposal to create the Advanced Campus of Catalão (photo 5) emerged through political means, with Professor Maria do Rosário Cassimiro and Professor Arédio Teixeira Duarte, both from Catalão, who held important positions within the Federal University of Goiás and were decisive influences on the local political scene (SILVA, 2005, not paginated, BALDINO, 1991, pp. 118-119).

Photo 5: Inauguration plaque for the Catalão/UFG Advanced Campus 1983
Author: CARNEIRO, V. A. (2008)

The two professors, Maria do Rosário and Arédio, with strong prestige and political representation within the University, were present and active in the history of the emergence of the Advanced Campus in Catalão (SILVA, 2005, not paginated). The Catalão Advanced Campus was inaugurated on 20 August 1983 and the Catalão City Council and UFG signed an agreement to set up two courses (CAC-UFG DIRECTOR'S REPORT, 2008, not paginated; SILVA, 2005, not paginated, DOURADO, 2001, p. 168).

Silva (2005, not paginated) emphasises the issue of agreements:

> According to the Internalisation Programme, human, material and financial resources would be made available not only by UFG, but also by State and Municipal Public Bodies and Private Bodies that signed mutual agreements with the University. With this in mind, UFG, through the Dean of Extension, has embarked on a policy of internalisation based on the signing of agreements with public and private bodies.

According to the CAC-UFG Board Report (2008, not paginated) and Dourado (2001, p. 158), the Catalão Campus of the Federal University of Goiás began its regular undergraduate teaching activities in 1986, when the first undergraduate programmes were set up.

Geography and Languages courses, through an agreement between the University and Catalão City Hall.

Even so, the Advanced Campus of Catalão was inaugurated on 17 December 1983 (CAC). This was the culmination of a political struggle that has borne excellent fruit for UFG and the entire Catalan community. The CAC was installed in the Teacher Training Centre, which was refurbished to function as an academic space and for extension activities (SILVA, 2005, not paginated). Silva (2005, not paginated) highlights these extension activities implemented in Catalão:

> The advanced campuses offered students a space in which they could carry out internships and also experience providing services to the local and regional community. In carrying out these activities, they generally had the support of their teachers and preceptors, the latter of whom were professionals from the towns with training in the respective areas and accompanied the university students during the period of their activities in the town and region. The emphasis given to teacher training indicates that on UFG campuses, in addition to internships and social assistance activities, degree courses would also be offered in order to meet a demand in Basic Education.

According to geographers Chaves et al. (2004, p. 134),

> In the mid-1980s, the Federal University of Goiás began a process of expansion and internalisation of its courses, mainly full degrees, and set up the Geography and Languages courses at the Catalão Campus under an agreement signed with the Catalão City Council. What was initially inaugurated to fulfil a temporary mission - training teachers to work in primary and secondary schools - ended up being strengthened by the struggles of teachers and students, until it became a UFG Teaching Unit, with seven degree courses and establishing itself as an important teaching and research hub in the south-east of Goiás.

The partnership between CAC/UFG and Catalão City Hall has led to the following the Literature and Geography programmes in 1986 (photo 6).

A few years later, new courses (implemented during 1988/2008) and other educational perspectives emerged at the CAC/UFG (photo 7) which, according to the CAC-UFG Board Report (2008, not paginated), showed that the UFG and the MEC were strengthening their relationship and led to the emergence of "two federal programmes: REUNI - the Programme for the Expansion and Restructuring of Public Universities (Federal Decree No. 6.096/2007) and the Programme for the Expansion of Federal Higher Education Institutions (created in 2005)". The aforementioned report (2008, not paginated) adds:

> The Catalão Campus was included in the category of New Campuses, accrediting it to receive investment in infrastructure, funding and federal places in the next selection processes (entrance exams).
> Through CONSUNI Resolution 19 of 11/11/2005, the Catalão Campus was transformed into a UFG Unit. The regulations were approved by CONSUNI at a meeting held on 23/11/2007 and instituted by Resolution No. 023 of 2007, which gives it the status of an off-site campus.

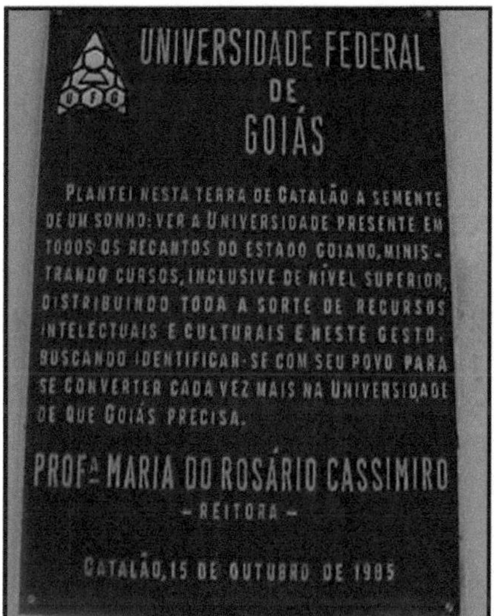

Photo 6: Plaque for the creation of the Letters and Geography courses on 15 October 1985
Author: CARNEIRO, V. A. (2008)

Photo 7: Partial view of the Advanced Campus of Catalão / UFG Source: SILVA, M. J. (2005)

Next, we'll tell you a little about the creation of the Geography programme at the Catalão Advanced Campus.

2.2.1.2. The creation of the Geography programme

The CAC/UFG Geography Course was set up in 1986 as part of the Higher Education Internalisation Project (1980) launched by the Federal University of Goiás, and was the first to be set up in south-eastern Goiás. Silva (2005, not paginated) interviewed Professor Elza Maria

Stacciarini (former Director and retired Professor of the Geography Course at CAC/UFG), who made the following comments:

> Teachers from the Geography and Languages programmes were in Catalão offering a training course to teachers from the state and municipal education systems. [...] These were courses that, if I'm not mistaken, the Office of the Dean of Extension was organising in agreement with the state and local councils. We stayed there for a fortnight. [...]. At the time we went to give this course, Professor Maria do Rosário Cassimiro was Pro-Rector of Extension, and we were asked to do a survey with the teachers about the interest in setting up higher education courses in that municipality. And there was a lot of interest. We applied a questionnaire not only to the teachers who were taking the course, but also to the community. And right at that time we went to see the facilities of the old Teacher Training Centre, which had an idea that the campus would be set up there. [...] This questionnaire showed a lot of requests for Geography, Portuguese and Biology [...].

This is how the first agreements to set up courses came about in 1985.

A degree in Literature and a degree in Geography.

Silva (2005, not paginated) continues to ask Professor Elza Maria Stacciarini about the process of setting up courses at the CAC, when asked about the choice of the first courses, she says:

> The choice of Geography and Letters I don't know if it came from the Rectorate, if it came from the City Council, if it was both in agreement. I do know that it wasn't the Geography Department or the Arts Department itself. It was a demand from the community and the Rector's Office took up the cause. The Rector's Office asked the Department to study the feasibility of setting up a course there. There were several meetings, so from March until September when it was approved, it was practically seven months of almost weekly meetings. In reality [...] there was no real gain in a meeting to implement the course. It was more in response to a request from the Rectorate and not because the Geography Department had approved it by consensus, unanimously or by an absolute majority. It was implemented because some professors committed themselves to taking the course [...].

Supporting the ideas of Silva (2005, not paginated),

> We can therefore say that the implementation of the first courses at the CAC was mainly the result of the political action of the Rector - under the management of Professor Maria do Rosário Cassimiro (a process that began when she was the Pro-Rector of
>
> Extension) and Mayor Haley Margon Vaz. We are not disregarding the participation of other segments of society, or even the trajectory of political leaders in the process of setting up a UFG Campus in the municipality, as well as the creation of higher education courses. The respective agreements between Catalão City Hall and the Federal University of Goiás commit to "mutual collaboration for the implementation" of Agreement No. 34/85 and Agreement No. 35/85 for Full Degrees in Literature and Geography.

The same author (2005, not paginated), with regard to Agreement No. 35/85 on the implementation of the Full Degree Course in Geography, presents the university's obligations:

> Clause Two - The University's Obligations
> 1. Offer the physical facilities of the Advanced Campus in Catalão for the course to run; 2. Allocate 50 (fifty) places from its entrance exams for the course in Catalão, to start in 1986; 3. Offer 2 (two) entrance exams: one for the course to start in 1986 and another in 1987. If there is demand and interest on the part of the parties to the agreement and the Geography Department is consulted, other entrance exams may be held; 4. To be responsible for the planning and didactic-pedagogical coordination of the course, through the Geosciences Collegiate and the Geography Department; 5. To carry out,

through its Undergraduate Pro-Rectory, the monitoring of the student's academic life; 6. Be responsible, through the Geography Department, for selecting the teaching staff to be hired by the City Council; 7. Train the selected teaching staff through refresher or specialisation courses in the field of Geography; 8. Be responsible for offering the course only within the period stipulated in the agreement. In this way, students who do not follow their class will be guaranteed the continuation of the course in Goiânia within the maximum time limit set by the Federal Education Council, if there is a vacancy and the student has justifiable reasons. The candidate will be duly informed of this rule when registering for the entrance exam; 9. be responsible for the transport, accommodation and meals of the University's lecturers and the Course Coordinator when on duty.

This scenario in Catalão, the scene of many discussions within the University, among Catalan politicians and demarches to the MEC, led to the installation of the Advanced Campus in Catalão. Silva (2005, not paginated) says:

> We're not going to say that the investments made by the public authorities in Catalão and the neighbouring town halls interfered in the decision made by the Rector's Office and the Departments of the Full Degree Courses in Geography and Letters, but in October 1985, UFG and the Catalão City Hall signed agreements to set up the respective courses. At the beginning of this process, the social actors who defined this policy of internalising courses were the Rector, the Pro-Rector of Extension and the Mayor of Catalão. At a later stage, the Departments of Languages and Geography joined them and, on the basis of what was agreed between them, the agreements and the guidelines and norms for the operation of the courses at the CAC/UFG were established.

According to the Political Pedagogical Project of the Geography Course/UFG (2005, p. 11),

> In 1986, the advanced campuses in the cities of Catalão and Jataí implemented specific classes for the undergraduate degree in Geography (BA and BSc), among others, as part of a policy to internalise UFG. These predominantly evening classes were linked to the curricular matrix of the Geography course at the headquarters, in the current IESA (then the Geography Department of the now defunct IQG-Institute of Chemistry and Geosciences), with distinct administrative and pedagogical specificities on each campus.

The project in question (2005, p. 12-13) describes:

> Geography, in its process of historical development as an area of knowledge, has been consolidating its theoretical and methodological position as a science that seeks to understand and explain the multiple interactions between society and nature. Thus, we must recognise that these transformations in the field of geographical knowledge have posed challenges to the training not only of the geographer-researcher, technician and planner, but also of the geographer-teacher in primary, secondary and higher education.

The aforementioned project (2005, p. 21-22) also emphasises this:

> The pedagogical dimension of the Geography degree programme will be developed. From the beginning of the course, it will be the responsibility of the Institute of Socio-Environmental Studies, the countryside campuses and the Faculty of Education at UFG, in view of the need to train professionals to work in basic education, without dissociating geographical knowledge, pedagogical practice and school content, in a systematic and continuous way.
>
> The structure of the degree reveals the need to develop the training of the geographer-educator, with mastery of geographical knowledge, school content to be socialised, establishing the link between its meanings in different contexts and its interdisciplinary dimension and, above all, the need to develop competences and skills relating to mastery of the didactic-pedagogical process.

Continuing along the lines of the Project (2005, p. 28-30), "the Licentiate qualification is aimed at students who intend to carry out professional teaching activities in Geography at the different levels and teaching modalities, which will have a workload of 2,984 hours to be studied in four academic years". In agreement with the Political Pedagogical Project of the Geography Course/UFG (2005, p. 39), it is emphasised that teaching, research and extension must be integrated:

> In turn, it should be included in everyday teaching, both as a time to apply spatial analysis techniques and as a way of enhancing students' ability to reflect on the reality in which they live. For geographers and geography, fieldwork is a traditional activity that should no longer be just a time for trips or excursions, usually to other places, and should no longer be restricted to a single subject. These activities, which continue to be important, should encourage interdisciplinary exchange, both within geographical knowledge and through the contribution of other knowledge.
> The field, therefore, should be the moment when research, teaching and extension merge in the knowledge of reality. In this sense, it must be an activity of reflection on research, as a technical professional and/or as a teacher, a less fragmented view of reality.

The aforementioned Project (2005, p. 51-52), via Resolution 730/2005 of CEPEC - UFG's Teaching, Research, Extension and Culture Council, sets out the full curriculum for the Bachelor's and Licentiate's Degree Courses in Geography (table 2), for students starting in the 2005 academic year and students who opt for this curriculum. And also in the use of its legal, statutory and regimental attributions, meeting in plenary session on 05/07/2005, in view of what is contained in process no. 23070007628/2005-21, resolves in Art. 7, item II, to define the modalities of fieldwork:

> **Fieldwork**: the Geography course will have two types of fieldwork to consider: a) fieldwork related to any subject in the course and focused on the specific content of a subject, which is the responsibility of the teacher, who will carry it out in accordance with the programme provided for in the course plan; b) interdisciplinary fieldwork carried out by a group of teachers in accordance with the provisions of the course plan for the respective subjects; c) fieldwork as a curricular component of a specific subject with a theoretical and practical workload (my emphasis).

The Project in question (2005, p. 52) emphasises that CEPEC/UFG Resolution No. 730/2005) in its Article 7, item III, states that the "practical activities of the Geography Course include different practical activities, both in the field and in the laboratory. These activities will be carried out individually or in groups of students, as provided for in the course plans of the subjects." The Geography Degree Course at the Advanced Campus of Catalão, Federal University of Goiás has a workload of 2,984 hours (table 1) for students to complete (CEPEC/UFG Resolution 730/2005), distributed as follows:

GENERAL DISTRIBUTION OF THE WORKLOAD OF THE GEOGRAPHY DEGREE COURSE	
Subjects	**Workload**
Common Core	1.472
Specific Core	928
Options	192
Free Core	192
Complementary Activities	200
Total	**2.984**

Table 1: General distribution of the workload of the Geography Degree Course Source: GEOGRAPHY COURSE PEDAGOGICAL POLICY PROJECT/2005

Organisation: CARNEIRO, V. A. (2007)

For a better clarification and breakdown of the general distribution of the Geography degree course's workload, see below:

1- Common Core: made up of compulsory subjects that guarantee training for professionals in Geography (bachelor's and licentiate's degrees), totalling 1,472 hours.

2- Specific Core: includes the subjects in the pedagogical area plus Supervised Internship I, II, III and IV, totalling 928 hours.

3- Optional subjects: in order to guarantee the continuous offer of optional subjects, the institution will opt for the system of offering them once every two consecutive academic semesters, whenever the number of students enrolled exceeds five, totalling 192 hours.

4- Free Core: consisting of at least 192 hours of a set of subjects freely chosen by the student, from among those offered for this core at any UFG academic unit.

5- Complementary Activities: participation in conferences, seminars, lectures, congresses, intensive courses, debates and other scientific, professional and cultural activities, including scientific initiation activities, totalling 200 hours.

Students regularly enrolled on the Geography Degree course at the Advanced Campus of Catalão at the Federal University of Goiás must complete the course with 2,984 hours.

The curriculum of the Geography degree course (table 2) offers subjects that are taught on a semester basis (1st year: 1st and 2nd semesters; 2nd year: 1st and 2nd semesters; 3rd year: 1st and 2nd semesters and 4th year: 1st and 2nd semesters).

When we looked through the teachers' diaries, we found notes on fieldwork in the urban perimeters, degraded areas, micro-basins, mining areas and rural areas of Catalão and surrounding municipalities. We also noticed that field activities are worked on in isolation by the component subjects of the curriculum in question (table 2).

The next step is to learn a little about the historical and geographical context of Pires do Rio in order to realise the establishment of a college with higher education courses in the areas of Literature, History and Geography.

CURRICULAR MATRIX FOR THE LICENCE IN GEOGRAPHY/CAC-UFG (CEPEC/UFG Resolution 730/2005)			
1st Semester		2° Semester	
Subjects	Workload	Subjects	Workload
Astronomy Fundamentals	**32**	**Thematic Cartography**	**64**
Basic Cartography	**64**	**Introduction to Climatology**	**64**
Geography and Society	**64**	**Forni, from Terril. and the Brazilian People**	**64**
Geography and Demography	**64**	**Population Geography**	**64**
Basic Statistics	**32**	**Geology and Mineral Resources**	**64**
Socio-spatial formation	**64**	**Free Core**	**32**
General Geology	**64**		
3° Semester		4° Semester	
Subjects	Workload	Subjects	Workload
Theory and Methodology of Geography	**32**	**Bemolo Sensing Principles**	**32**
General Geomorphology	**64**	**Geoprocessing**	**32**

Dynamic Climatology	64	Theory and Methodology of Contemporary Geography	64
Geopolitics and Political Geography	64	Agricultural Geography	64
Geography of Goiás (oplaliva)	64	Free Core	64
Philosophical and Philosophical and Socio-Historical Foundations of Education	64	Tropical Geomorphology (oplaliva)	64
Geography and Education (oplaliva)	64	Theory and Practice of Planning (oplaliva)	64
Fundamentals of Environmental Education	32	Geopolitics and Geog. Politics (oplaliva)	64
Cultural Geography (oplaliva)	32	Begion Theory and Regionalisation (oplaliva)	32
		Didactics and Teacher Training	32
		Educational Psychology I	64
		Geography, Subject and Culture (oplaliva)	32
5° Semester		6° Semester	
Subjects	Workload	Subjects	Workload
Pedology	64	Geography of Industry	64
Urban Geography	64	Research Methodology	32
Free Core	64	Teaching Geography I	64
Supervised Internship I	96	Terrilorio and Bedes (oplaliva)	64
Climatology Applied to Geography (oplaliva)	64	Educational Policies in Brazil	64
Didactics and Geography Teacher Training	64	Biogeography (oplaliva)	64
Educational Psychology II	64	River Basin Analysis and Management (oplaliva)	64
Regional Shields: Laline America (oplaliva)	64	Environmental Planning (oplaliva)	64
Geography and Social Movements in the Countryside (oplaliva)	64	Supervised Internship in Geography II	96
Environmental Education Melodies and Practices (oplaliva)	32		
7° Semester		8° Semester	
Subjects	Workload	Subjects	Workload
Preparing a Research Project	64	Final Course Work	64
Teaching Geography II	64	Free Core	64
Territorial Planning (oplaliva)	64	Supervised Internship in Geography IV	64
Ecology of the Cerrado (oplaliva)	64		
Culture. Curriculum and Assessment (oplaliva)	64		
Supervised Internship in Geography III	160		
Geography and Tourism Planning (oplaliva)	64		
Overall workload: 2,984 hours			

Chart 2: Curriculum matrix for the Degree in Geography/CAC-UFG
Source: PEDAGOGICAL POLICY PROJECT OF THE GEOGRAPHY COURSE/UFG, 2005
Organisation: CARNEIRO, V. A. (2007)

2.2.2. Getting to know Pires do Rio

The municipality of Pires do Rio (map 3) is located in south-eastern Goiás and is part of the Pires do Rio Geographical Microregion (figure 5). Its origins are associated with the implementation of the Goiás Railway during the first decades of the 20th century, under the government of the then President of the Republic Epitácio Pessoa.

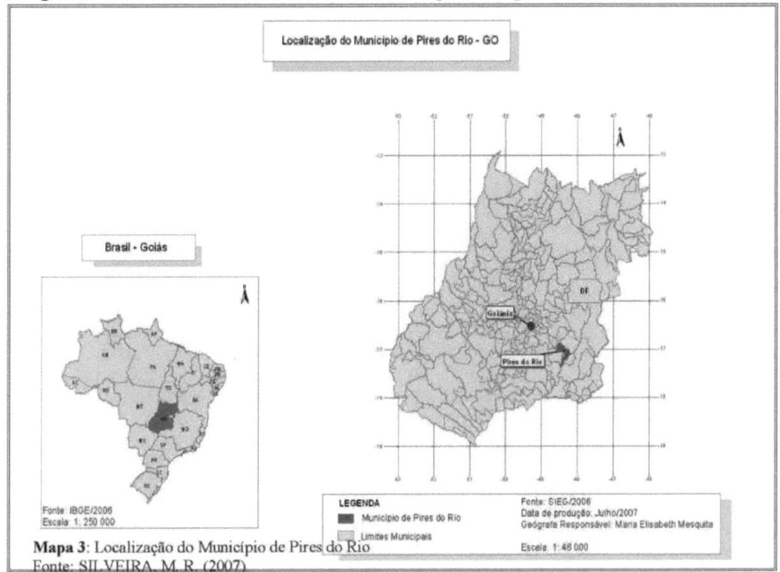

Mapa 3: Localização do Município de Pires do Rio
Fonte: SILVEIRA, M. R. (2007)

Map 3: Location of the Municipality of PireS do Rio Source: SILVEIRA, M R(2007)

Municípios: Gameleira de Goiás, Silvânia, Vianópolis, Orizona, São Miguel do Passa Quatro, Cristianópolis, Santa Cruz de Goiás, Palmelo, **Pires do Rio** e Urutaí.

Figure 5: Pires do Rio micro-region
Source: SEPLAN/SEPIN (STATE OF GOIÁS), 2007
Organisation: CARNEIRO, V. A. (2007)

Silva (2006, p. 97) explains that:

> With the development of capitalism in Brazil at the beginning of the 20th century, the spatial configuration of the hinterland changed. Goiás became part of the national market through the expansion of the frontier and the emergence of new transport routes, such as the construction of the railway. With the railway, new settlements were formed, such as Pires do Rio.

Silveira (2007, p. 19), Borges (1990, p. 17), Gomes and Teixeira Neto (1993, p. 75) and Ferreira (2005, p. 62) state that "the municipality of Pires do Rio arose as a result of the implementation of the railway in the south-eastern part of the state of Goiás".

In the 1920s, engineer Balduíno Ernesto de Almeida, director of the Goiás Railway, based in Araguari (state of Minas Gerais), decided to set up a railway station on the other side of the Corumbá River (right bank), as the construction of the Epitácio Pessoa Metallic Bridge (a tribute to the President of the Republic at the time) would enable the railway to continue to Anápolis (state of Goiás).

According to Ferreira (2005, p. 59), the creation of the Companhia Estrada de Ferro Goiás made it possible:

> After several changes to the original railway project, the actual construction of the line began on 23 December 1909, marked by the construction of the Araguari Railway Station, which was to reach the city of Goiás, the state capital. In 1912, the tracks crossed the Paranaíba River and penetrated the south of Goiás. The slow trajectory of the Goiás Railway consisted of two stages. The first, the Araguari-Roncador section (233 kilometres of track including the 23-kilometre Goiandira-Catalão branch), was built between 1909 and 1914, a relatively short period considering the difficult topography of the region. The second, the Roncador-Anápolis stretch, took longer, from 1922 to 1935. For almost a decade the road was paralysed in Roncador, on the banks of the Corumbá River, technically prevented from crossing the great river.

According to Paes (1991, p. 19), the town of "Pires do Rio was founded on 9 November 1922. This was the date on which the Epitácio Pessoa Metallic Bridge and Railway Station was inaugurated, over the Corumbá River, 120 metres long, considered at the time to be a monumental engineering feat". In order for this historic event to take place, the founding of Pires do Rio, Colonel Lino Teixeira Sampaio (owner of the Brejo Farm), in agreement with the board of the Goiás Railway, donated a plot of land so that the railway station, a school and a church could be built (SIQUEIRA, 2006, p. 28-30). The small village, which had only one central street, Rua do Fogo, quickly prospered and became the town of Pires do Rio (PAES, 1991, p. 19-20; SILVA, 2006, p. 100; PREFEITURA MUNICIPAL DE PIRES DO RIO, 1997, p. 2).

CODEPLAN - Companhia de Desenvolvimento do Planalto Central (1982, p. 3), in its report studying the potential of the municipalities in the geo-economic region of Brasília, emphasises that:

> The town that later became the city of Pires do Rio arose as a result of the establishment of the "Goiás Railway". Around 1920, the director of the railway, engineer Balduíno Ernesto de Almeida, proposed creating a town next to the railway's first station, for which he donated three bushels of land on the "Brejo" farm, owned by Mr Lino Teixeira de Sampaio.

Paes (1991, p. 20) says:

> A population centre began to emerge around the railway station, which was designed by Álvaro Sérgio Pacca, chief designer of the Goiás Railway, to whom Engineer Balduíno allegedly said: "YOU WILL SEE WHAT A CITY THIS WILL BE IN TWENTY YEARS". And so our young city, which had been emerging, had its foundation consecrated, and now all that was left to do was to grow, and grow, and this happened quickly.

The town of Pires do Rio was named in honour of the Minister of Transport and Public Works José Pires do Rio (a member of the senior staff of the President of the Republic

Epitácio Pessoa), who was in south-eastern Goiás in 1921, inspecting the construction of the Metallic Bridge (photo 8) and the expansion of the railway section over the Corumbá River. Silveira (2007, p. 52) and Ferreira (2005, p. 59) report that "due to technical difficulties, especially when crossing rivers, at that time the Corumbá River, the tracks were stopped at Roncador in 1914".

During this short period of time, with the stoppage of work on the Roncador (1914), the technical problem of crossing the Corumbá River was solved; from the Roncador side (left bank) to the side of the future city of Pires do Rio in 1922 (right bank), noting that the railway works were reactivated in 1919 by order of the Presidency of the Republic at the time.

Photo 8: Epitácio Pessoa metal bridge over the Corumbá River Source: Pires do Rio City Hall (1997)

The town and railway station of Roncador, located on the left bank of the The Corumbá River "lost" its economic significance as an entrepôt, fell into decay and "disappeared" after the extension of the railway and the construction of the Epitácio Pessoa Metallic Bridge over the aforementioned river, reaching the right bank and future railway continuity.

Ferreira (2005, p. 69) says that:

> The people of Roncador, aware of the realities of the railway and foreseeing the loss of the privilege they enjoyed as a town at the terminus of the tracks, the sole reason for their existence, began to abandon the place in search of another "commercial point". Roncador, it was certain, would disappear as quickly as it was born and grew.
>
> This fact constituted the phenomenon known here as "phagocytosis", which was essential, at first, for the formation of the town of Pires do Rio. Roncador was extinguished to create the nascent town.

In Nogueira (1977, 40),

> The second stage of the construction of the Goiás Railway was marked by two simultaneous events that were immediately important for the region, as we shall see: the inauguration of the Epitácio Pessoa Bridge over the Corumbá River and the inauguration of a railway station that would consolidate the newly created town of Pires do Rio.

The population of Povoado do Roncador, saddened by the stagnation of the area, soon tried to incorporate itself into the newly created town of Pires do Rio. For a long time, the Goiás Railway ended at the Roncador stop (on the left bank of the Corumbá River), which prevented the railway from being extended. It was a geographical obstacle that could only be overcome if a metal bridge was built there, which would allow the railway to reach the former state capital, now the city of Goiás.

Borges (1990, p. 107) explains the reason for the "withering away" of Povoado do Roncador in south-eastern Goiás:

> Roncador was the last station built before the Goiás Railway Company was taken over by the Federal Governor. Inaugurated in 1914, unlike others that became towns and municipalities, Roncador, after being one of the largest commercial centres along the railway, fell into decay. With the construction of the Epitácio Pessoa Bridge over the Corumbá River in 1922 and the extension of the railway line, the town fell into crisis until it disappeared completely.

Nogueira (1977, p. 44) also highlights the issue of Povoado do Roncador:

> The people of Roncador, experienced in the daily struggle of commerce, were attentive to the realities of their environment. In daily contact with passers-by and railway workers, especially the chief engineers, they kept well informed of the programme to lay the tracks. She was able to analyse the consequences well. She foresaw the end of the lively village she had built. She clearly understood that if she lost the privilege she enjoyed of being a railway terminal, Roncador would lose the sole reason for its existence.

Corroborating the ideas of Borges (1990, p. 110),

> It is undeniable that the Goiás Railway was the main modernising agent that began and accelerated the process of transforming the structures of regional society. However, one should not fall into theoretical simplifications and hasty analyses, attributing to the railway the role of the only factor in socio-economic and cultural change. As well as not being the only modernising element, the railway itself is not an autonomous phenomenon that promotes transformations, since it itself was the result of Latin American modernisation. What is correct is to consider the railway as one of the main instruments built and used by capital in the modernisation and incorporation of the agrarian economy into the domains of imperialism.

From its beginnings to the present day, the municipality of Pires do Rio (photos 9, 10 and 11) has an economy geared towards the primary sector (farming and cattle-raising). Researcher Ferreira (2005, p. 60) shows that:

> The penetration of the railroad into Goiás provoked a significant transformation in the south and southeast regions, with the modernisation of the agrarian economy. At the same time, other aspects of social, political and cultural life also began to change, giving the population a sense of "urbanity". There was an awakening from an era of isolation with a strictly rural economic and ideological predominance, marked by subsistence production and extensive livestock farming, in which settlements were made by the formation of farms that then became urban "patrimonies".

The same author (2005, p. 62-63) states that:

> In the 1940s and 1950s, linked to the benefits of the railway, Pires do Rio was an important city in Goiás, standing out not only for its economy, but also, and above all, for its urban consolidation on a modernising basis for the regional context. These benefits continued until the 1960s, despite the frequent and growing crises of the Goiás Railway and, in general, of Brazilian railways, resulting above all from the supremacy of the road transport system in the country.

In the 1980s, the arrival of soya on the plateaus of south-eastern Goiás changed and boosted the economic picture of the area mentioned above and Silveira (2007, p. 62) reports that:

> This situation led the municipality of Pires do Rio to consolidate its public and private infrastructure, meeting the demands of the region. This has led to the installation of large-scale agro-industries. OLVEGO - Óleos Vegetais Goiás (works with soya crushing), in

> the early 1990s, and NUTRIZA AGROINDUSTRIAL (works with chicken slaughter) in the middle of the same decade. With the establishment of these agro-industries - which belong to the TOMAZZINI GROUP - the municipality grew economically stronger, attracting new companies and consolidating its commercial infrastructure.

The same author (2007, p. 63-64) explains that "the company NUTRIZA AGROINDUSTRIAL, which belongs to the TOMAZZINI GROUP, works with two brands: NUTRIZA and FRIATO (frozen chicken cuts), which drive the economy of Pires do Rio and the surrounding area."

Photo 9: Aerial view of Pires do Rio
Source: Prefeitura Municipal de Pires do Rio (1997)

Photo 10: Panoramic view of Pires do Rio Author: CARNEIRO, V. A. (2008)

Photo 11: Arrival in Pires do Rio via the GO-020 motorway Author: CARNEIRO, V. A. (2008)

In a recent study of the area of influence of Pires do Rio, Silveira (2007, p. 14) outlined "a regional profile for the municipality of Pires do Rio, coining the name 'TRILHO DAS PENAS' Region, due to the agro-industrial issue (poultry sector)".This geographical and historical horizon led to the establishment of a state college (photo 12) in Pires do Rio (1990s), via the decentralisation of higher education in Goiás, promoted by the Goiás State Department of Education and Culture/Superintendence of Higher Education.

Photo 12: Aerial view of the State College in Pires do Rio - 1994 Source: State University of Goiás website (2007)

In this way, my research will focus on the Geography Degree Course, which today belongs to the state college in Pires do Rio (set up in the 1990s) with its teachers and students.

The next item will deal with the issue of the creation of the Pires do Rio University Unit of the State University of Goiás.

2.2.2.1. The creation of the Pires do Rio University Unit

The University Unit of Pires do Rio, before becoming an integral part of UEG - State University of Goiás, according to "State Bill No. 245 of 27/06/1985, was called "Faculty of Sciences and Letters of Pires do Rio" (BALDINO, 1991, p. 202).

State deputy João Natal (a PMDB politician, in memoriam) presented the aforementioned bill with the following justification for setting up a university centre in Pires do Rio:

> Among the reasons that militate in favour of the creation of the Faculty of Sciences and Letters in the traditional and cultured city of Pires do Rio, the one that refers to the state's obligation to provide young people from the interior with the means and conditions to complete their higher education in the land that served as their birthplace, thus avoiding their costly and often uncomfortable journey to large urban centres, is of the utmost importance.
>
> It is worth noting that the beautiful and progressive city of Pires do Rio has been experiencing an extraordinary phase of socio-economic development, with undeniable repercussions in the various fields of human activity.

The bill presented by State Representative João Natal was considered by the plenary and obtained a favourable result for the creation of a college in Pires do Rio (BALDINO, 1991, p. 202). For this case and many others about the expansion of higher education in the interior of Goiás, we endorse Ferreira's idea (2006, p. 66) that "the 1980s in Goiás were prodigious in linking higher education with political-electoral agreements, as a political marketing strategy for candidates in electoral campaigns or to consolidate their populist policies".

Ferreira (2006, p. 66) adds that "during this period, several institutions were created in development pole cities such as Anápolis, Jataí, Rio Verde, Catalão, Itumbiara, Porangatu, Luziânia, among others". Thus, on 14 October 1985, the Goiás State Legislative Assembly, via State Law No. 9,805, authorised the creation of the Faculty of Education, Sciences and Letters of Pires do Rio. The change of name from Faculty of Sciences and Letters to Faculty of Education, Sciences and Letters did not cause any problems, as the fight behind the scenes in the state legislature was victorious (UEG, 2008, not paginated).

On 16 April 1986, State Decree No. 2,577 created the Faculty of Education, Sciences and Letters of Pires do Rio as an autarchy. The Higher Education Institution in Pires do Rio was created in 1985 and the next step was to get it up and running with authorisation from the President of the Republic of Brazil (UEG, 2008, not paginated).

In Goiás, the State Legislative Assembly replaced the name Faculdade de Educação, Ciências e Letras de Pires do Rio (Faculty of Education, Sciences and Letters of Pires do Rio) with Faculdade Estadual Celso Inocêncio de Oliveira (UEG, 2007, not paginated). It's worth remembering that this institution, the one in Pires do Rio, remained on paper for almost nine years and had no headquarters. However, according to researchers Ferreira (2006, p. 66) and Baldino (1991, p. 66-121), most of the state initiatives (both legislative and executive) remained on paper, i.e. laws were created for the creation of isolated colleges, but they were not implemented. It wasn't until 4 March 1994 that a Presidential Decree (figure 6) authorised the Celso Inocêncio de Oliveira College in Pires do Rio, in the state of Goiás, to run courses in Literature, Geography and History (UEG, 2007, not paginated).

The Celso Inocêncio de Oliveira State College, based in the municipality of Pires do Rio, is a state higher education establishment, created by State Decree No. 2.577 of 16/04/1986 - Official Gazette of 28/04/1986, in accordance with the legislative authorisation contained in State Law No. 9.805 of 14/10/1985, published in the Official Gazette of 25/10/1985 (UEG, 2008, not paginated).

According to Decree no. 9.805, of 14 October 1985, "the Higher Education Institution was a State Autarchy, linked to the Goiás State Department of Education, Culture and Sport". The aforementioned Higher Education Establishment was a body linked to the indirect administration of the State Executive Power, enjoying all the advantages guaranteed to Public Law entities, including the issue of patrimonial, financial, administrative, disciplinary and didactic-pedagogical autonomy, observing the principles of jurisdictional dependence in relation to the direct administration of the State of Goiás (UEG, 2008, not paginated).

The Celso Inocêncio de Oliveira State College, known to the Piresinos and the surrounding area as FAESCI (figure 7), in this period, according to State Decree No. 2.577/1986, aimed to train teachers and produce research and extension through the Degree Courses in Geography, History and Letters (UEG, 2007, not paginated).

From my own experience, research didn't take place at FAESCI and extension was restricted to just a few domestic events (lectures, mini-courses and pedagogical weeks for courses on the institution's premises). According to Baldino (1991, p. 210-211), the central concern of FAESCI, which was linked to the State Department of Education and Culture/Superintendence of Higher Education, was always to train teachers for the educational market (primary and secondary education) in the Pires do Rio micro-region. Ferreira (2006, p. 18) explains that:

> Before the creation of UEG, the state colleges in Goiás had little or no relationship with each other. At the time, there was no concern about establishing a single database, for example, or unifying curricula, timetables, methodologies, etc. When UEG was created, it was also necessary to establish a communication channel that would make it possible to bring together data from the various university units that made up UEG. For this reason, the institution's data (number of students, teachers, staff, courses, budgets, extension and research programmes, among many others) only began to be more systematised after 2002, when a more effective internal communication policy began.

During the administration of Goiás State Governor Marconi Perillo, UEG was created - Universidade Estadual de Goiás (Goiás State University) - by means of State Law No. 13,456 of 16 April 1999. Its implementation ended up undertaking public policies for the expansion and internalisation of State Higher Education in Goiás (UEG, 2007, not paginated).

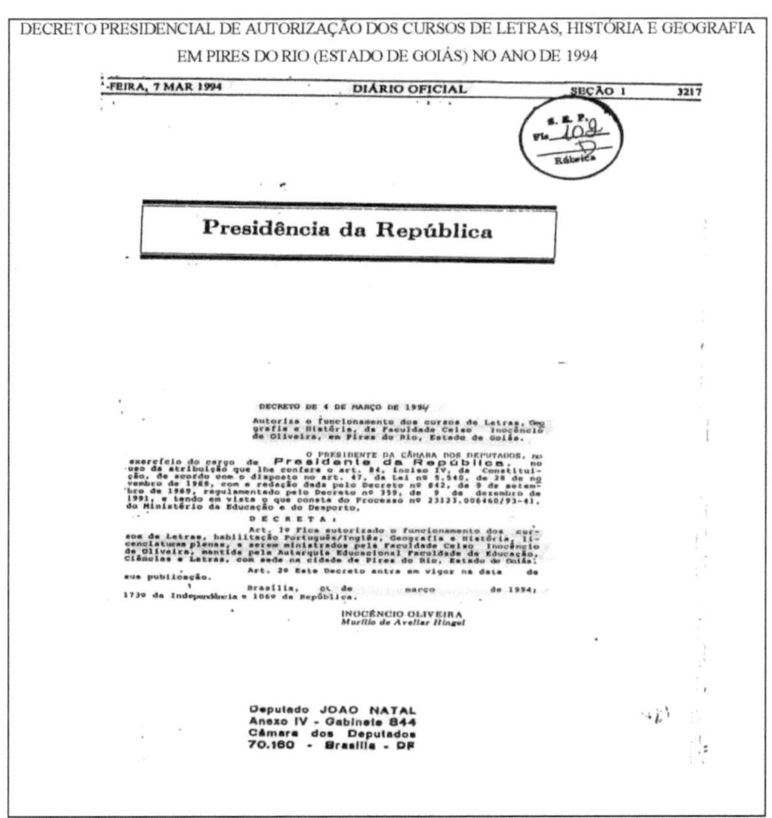

DECRETO PRESIDENCIAL DE AUTORIZAÇÃO DOS CURSOS DE LETRAS, HISTÓRIA E GEOGRAFIA EM PIRES DO RIO (ESTADO DE GOIÁS) NO ANO DE 1994

-FEIRA, 7 MAR 1994 DIÁRIO OFICIAL SEÇÃO I 3217

S. E. P.
Fls. 102
Rúbrica

Presidência da República

DECRETO DE 4 DE MARÇO DE 1994

Autoriza o funcionamento dos cursos de Letras, Geografia e História, da Faculdade Celso Inocêncio de Oliveira, em Pires do Rio, Estado de Goiás.

O PRESIDENTE DA CÂMARA DOS DEPUTADOS, no exercício do cargo de **Presidente da República**, no uso da atribuição que lhe confere o art. 84, inciso IV, da Constituição, de acordo com o disposto no art. 47, da Lei nº 5.540, de 28 de novembro de 1968, com a redação dada pelo Decreto nº 842, de 9 de setembro de 1969, regulamentado pelo Decreto nº 359, de 9 de dezembro de 1991, e tendo em vista o que consta do Processo nº 23123.006460/93-41, do Ministério da Educação e do Desporto,

DECRETA:

Art. 1º Fica autorizado o funcionamento dos cursos de Letras, habilitação Português/Inglês, Geografia e História, licenciaturas plenas, a serem ministrados pela Faculdade Celso Inocêncio de Oliveira, mantida pela Autarquia Educacional Faculdade de Educação, Ciências e Letras, com sede na cidade de Pires do Rio, Estado de Goiás.

Art. 2º Este Decreto entra em vigor na data de sua publicação.

Brasília, 04 de março de 1994; 173º da Independência e 106º da República.

INOCÊNCIO OLIVEIRA
Murílio de Avellar Hingel

Deputado JOAO NATAL
Anexo IV - Gabinete 844
Câmara dos Deputados
70.160 - Brasília - DF

Figure 6: Presidential decree authorising Literature, History and Geography courses in Pires do Rio (Goiás state) in 1994
Source: COORDINATION OF THE GEOGRAPHY COURSE/UEG-PIRES DO RIO, 2007
Organisation: CARNEIRO, V. A. (2007)

Figura 7: T-shirt print from the Geography class of 1994 Source: CARNEIRO, V. A. (1996)

According to Ferreira (2006, p. 56),

> The creation of UEG in 1999 can be considered a milestone in the history of higher education in Goiás, because at a time when the private sector was largely responsible for the expansion of higher education in Goiás and Brazil, a public university emerged after decades of demands for its creation.

In 2000, FAESCI - Celso Inocêncio de Oliveira State College - was renamed the Pires do Rio University Unit of UEG - Goiás State University. In this way, FAESCI was added to the government's "New Time" project for Goiás, in the letter of proposal for higher education, coining the foundation of the UEG of the then Marconi Perillo state government, which was successful in the election.

According to State Law No. 13.456 of 16 April 1999, published in the State Official Gazette on 20 April 1999, UEG's mission is "to produce and socialise scientific knowledge and know-how, to develop culture and the integral formation of professionals and individuals capable of critically integrating themselves into society and promoting the transformation of the socio-economic reality of the state of Goiás and Brazil" (UEG, 2008, not paginated).

According to Tavares and Marinho (2006, p. 33-34),

> Public education policies in Goiás are in line with the policies developed by the federal government, which in turn are based on international guidelines. In the mid-1990s, there was no development project for public higher education in Goiás, and basic education was prioritised, which led to the expansion of private higher education.
> This situation changed in the final years of the last decade when the State University of Goiás was created by then Governor Marconi Ferreira Perillo Júnior through Law No. 13,456 of 16 April 1999. The establishment of UEG ended up implementing public policies for the expansion and internalisation of education.

> State Higher Education in Goiás, even though it goes against current practice and the demands of public policies defined by neoliberalism, which advocates the privatisation of this level of education, as well as the interests of private higher education advocates.
> The movement for the creation of a public, free and quality university in Goiás is not a recent proposal when you consider that mobilisation for the creation of this institution has existed since the 1950s.
> In line with Law 5.540 of 2 November 1968, which deals with university reform and allows for the fragmentation of higher education, isolated colleges were created in Goiás in the 1960s. Various initiatives fuelled the expansion of higher education, whether by the private sector or by federal, state or municipal authorities in the 1960s and 1970s and, more markedly, in the last two decades of the 20th century. At the time, several higher education autarchies were created by the state government in the form of Faculties of Education, Sciences and Letters.
> The social movements of both teachers and students in favour of the internalisation of higher education became evident through the 1st and 2nd SEMINARS ON THE EXPANSION OF 3rd GRADE EDUCATION, organised in 1986 and 1987 by the Regional Delegation of the Ministry of Education and Culture in Goiás (DEMEC).
> Also, in the 1980s, the State Students' Union (UEE) debated and fought for the creation of a multi-campus university, which had its origins in the Goiás School of Physical Education (ESEFEGO), the Anápolis Faculty of Economic Sciences (FACEA) and was later transformed into the Anápolis State University (UNIANA), with Law No. 11,655/1991, and generated expectation around the possibility of unifying the isolated state faculties created by the governors of recent decades.

Researcher Silva (2005, p. 96) adds:

> In Goiás, this consolidation culminated in the creation of UEG, a banner that emerged from intense debates promoted by the university student movement, through the State Student Union (UEE), and political alliances and agreements with the Goiás state government.

The creation of UEG, therefore, results from the process of transforming UNIANA (formerly FACEA-Faculdade de Ciências Econômicas de Anápolis) and incorporating the 13 isolated State Higher Education Institutions (figure 8), which does not happen at any given time, but is consolidated as a campaign promise by the candidate for State Government, Marconi Perillo, who receives the support of teachers and students from the state student movement (TAVARES; MARINHO, 2006, p. 33-34; FERREIRA, 2006, p. 60-65, SILVA, 2005, p. 94-95).

Silva (2005, p. 95-96), mentions that:

> In the 1980s, under the Iris Rezende Machado (1983-1986) and Henrique Santillo (1987-1991) governments, public higher education received a new boost. During this period, isolated colleges were created in the interior of the state, responding to local demands, but with a conservative modernisation policy, given the form of management to which they were subjected, some operating in borrowed buildings, without libraries and without official recognition of many of the courses on offer. It is important to emphasise that many of the colleges created were not put into operation and only came into being in 1999, under the new government, as part of the group that formed the State University of Goiás - UEG.

According to Ferreira (2006, p.60-65),

> The state's higher education system, which until 1999 was made up of 13 isolated institutions, with the creation of UEG, they are amalgamated into a single institution that expands to 30 University Units and 20 University Centres, spread across all regions of Goiás.

UEG is an institution that has a rather peculiar characteristic in relation to other higher education institutions: it is multicampi (SILVA, 2005, p. 92; PERINI, 2006, p. 76), i.e. it is territorially decentralised (figure 9). Professor Rachel Teixeira (State Education Secretary in Marconi Perillo's government), interviewed by Ferreira (2006, p. 80-81), states that there were two controversial issues that preceded the process of creating UEG. There was no consensus among members of the Goiás state government on the model to be adopted for the state institution:

> The idea was to create two institutions: The first, located in Anápolis, which would be geared towards technological innovation and research development (neo-Humboldtian model) while the other isolated colleges could be united in a university centre, which would be geared towards teacher training (Napoleonic model), which, as said before, had a very high demand in the state, since only 32% of teachers in the state network had higher education training.

To this question, Ferreira (2006, p. 81), adds:

> Another factor was the resistance of the academic communities of UNIANA in Anápolis and ESEFEGO in Goiânia, who differed on the possible format of the future university. The two academic communities (the largest in what was then the future UEG) did not want to lose their status as both a university and a college, which they had historically consolidated.
>
> The commission responsible for the debates, headed by the Office of SECTEC - the Secretariat for Science and Technology, presented the guidelines for the work before the enactment of State Law No. 13,456 of 16 April 1999, which created UEG from the transformation of UNIANA and the amalgamation of the isolated colleges in the interior and the state capital.
>
> Soon after the law creating UEG was published, the first controversy arose: in which university unit would the new university be based?
>
> This controversy was sparked by the teachers, staff and students of ESEFEGO, who did not accept that UEG's headquarters should be located outside the city of Goiânia, either because of the school's tradition or because of political issues.

Professor Rachel Teixeira, continuing her interview with Ferreira (2006, 82), comments:

> In the first meetings to organise the UEG, in addition to these issues, the political position of these professors, many of them linked to left-wing parties, who should, on principle, oppose the Governor, was made explicit. But the Governor was unyielding and since Anápolis, the then candidate, had been elected with more than 75 per cent of the votes, it would be a reward to the city to give it the headquarters of UEG.

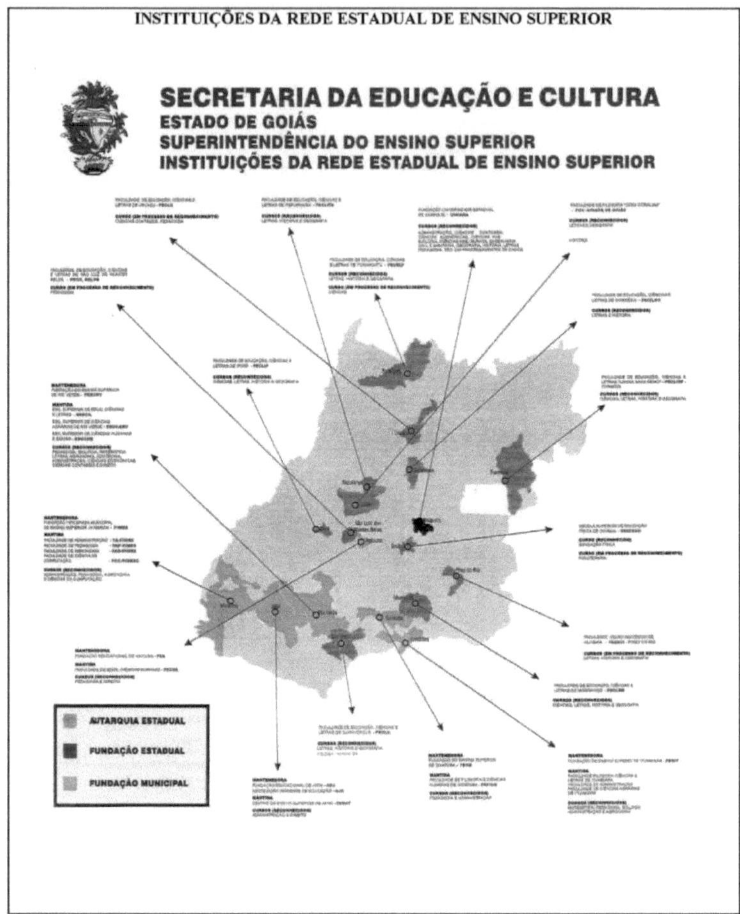

Figura 8: State higher education institutions
Source: Geography Course Coordination/UEG-Pires do Rio, 2000
Organisation: CARNEIRO, V. A. (2007)

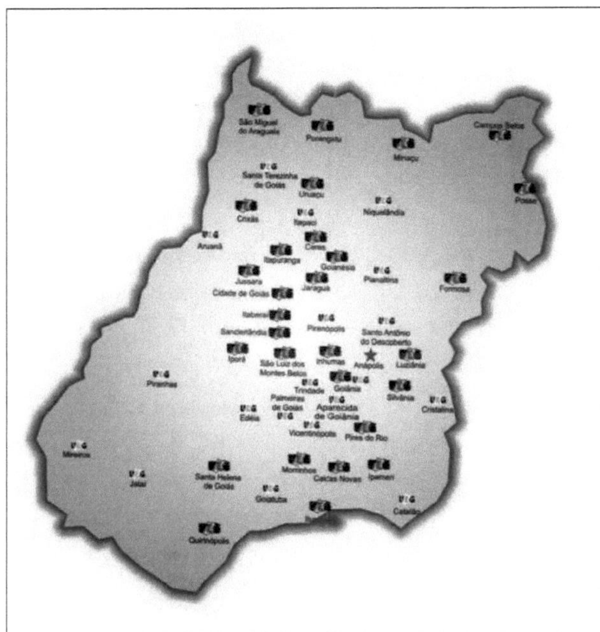

Figura 9: Geographical location of UEG's university units and centres
Source: Pró-Reitoria de Graduação/UEG, 2006

We agree with Ferreira (2006, p. 83-91),

> In these early days of UEG, the years of disintegration of the isolated faculties ended up being an obstacle to overcome, because each one had a different purpose (when it did), different projects (or not), in short, each University Unit that formed UEG needed to incorporate the "university spirit".
> Since its creation, UEG has had a presence in the various regions of the state, through its University Units, University Centres and Extension Courses. This process, extremely important for the consolidation of UEG as a public institution, also caused, and still causes, great concern, since the need for quantitative expansion had to be accompanied by quality.

State Law No. 13.456/1999 organically links UEG to the State Department of Education and, later, to the Department of Science and Technology (SECTEC), through State Decree No. 5.158/1999 (UEG, 2007, not paginated). As a multi-campus organisation, UEG runs ten degree courses in Geography in the following municipalities in Goiás: Formosa, Goiás, Itapuranga, Iporá, Minaçu, Morrinhos, Pires do Rio, Porangatu and Quirinópolis (in the evening) and Anápolis (in the morning).

Ferreira (2006, p. 91) emphasises that:

> UEG's initial expansion was basically through degree courses (Pedagogy, History, Geography, Languages and Maths) in order to meet the repressed social demand, as well as the need to improve the qualifications of basic education teachers in Goiás. Although necessary, the institution itself recognises that the expansion in the number of courses and University Units (UnU) needed basic criteria in order to be carried out.

This issue is crucial, because in the corridors of the head office in Anápolis and the university units in the interior of the state, the discussion that UEG is expanding at the behest of politicians often comes up among teachers, administrators and students, and the issue of teaching quality takes a back seat. Researcher Ferreira (2006, p. 91) says that "in addition to the real need to expand UEG, the issue of political and electoral influences was also a determining factor".

In an interview granted to the same researcher (2006, p. 91), Professor Gilvane Felipe (Goiás State Secretary for Science and Technology) says, "Every week a mayor, a member of parliament, a councillor would turn up and ask for UEG units to be opened in certain municipalities. The Rector, at the time, granted a number of these requests. That's when I realised that his interest was clear: to run for Governor."

Ferreira (2006, p. 105) goes on to say: "It is clear that an expansionist process of the magnitude that has taken place at UEG has not been accompanied by all the necessary indicators so that it can fulfil the minimum parameters of academic quality and social quality".

We can therefore say that UEG, in its dialectical process of socialisation and production of knowledge, establishes itself as part of a political practice that, by producing something new, meets the real needs of a given population. However, it is from the union of the isolated faculties into a University that the basic structure of teaching, research and extension is established (TAVARES; MARINHO, 2006, p. 41).

According to Silva (2005, p. 93),

> The state public university of Goiás aims to be a space for the production of knowledge, a locus for human and professional training. As a social and educational institution, its principles are based on the ability to seek knowledge and build knowledge.

UEG also seeks to articulate itself within the framework of the sustainable development project for the state of Goiás, endeavouring to promote itself as an institution that is not just a reproducer of knowledge, but as a producer of knowledge (TAVARES; MARINHO, 2006, p. 41).

Next, we'll find out about the creation of the Geography Degree Course at the University of Lisbon.

Pires do Rio University Unit of the State University of Goiás.

2.2.2.2 The creation of the Geography programme

It wasn't until the Presidential Decree of 4th March 1994 (UEG, 2007, not paginated) that the degree courses in Literature, Geography and History were actually authorised to begin at the former Celso Inocêncio de Oliveira-FAESCI State College, now the Pires do Rio University

Unit of the UEG-Goiás State University (photo 13).

For Tavares and Marinho (p. 56-57),

> The aim of setting up a higher education centre in the municipality of Pires do Rio was to train and improve teachers in the region. To this end, we considered, above all, the lack of a higher education course in the region and the difficulty in moving the population to larger centres in search of higher education. The need to meet the demand for higher education in the following municipalities was also taken into account: Bela Vista de Goiás, Caldas Novas, Morrinhos, Campo Alegre, Cumari, Cristalina, Goiandira, Ipameri, Orizona, Palmelo, Piracanjuba, Pontalina, Santa Cruz de Goiás, Silvânia, Urutaí, Vianópolis, São Miguel do Passa Quatro, Cristianópolis and others.

The UEG-Pires do Rio Academic Guide (2006, p. 1), in its initial message, states:

> The State University of Goiás, present in Pires do Rio since 1994 as the Celso Inocêncio de Oliveira State College (FAESCI), represents a path towards the internalisation of higher education and one of the recognised symbols of growth and development for the people who live in the region. It also represents the aspirations of all its staff, students, partners, the community, in short, of countless people who look to it as an opportunity to realise their dreams.

Photo 13: UEG-Pires do Rio façade
Source: CARNEIRO, V. A. (2008)

Going back a little in time, the construction of the college building began on 20 October 1992 and was followed by the name change from Faculty of Education, Sciences and Letters (State Law No. 9.805/1985 and State Decree No. 2.577/1986) to Celso Inocêncio de Oliveira Faculty-FAESCI (State Law No. 11.901/1993).

The college is named after Mr Celso Inocêncio de Oliveira (an influential lawyer in the Railway Region and a member of the PMDB), as he was one of the people involved in the process of setting up the institution in Pires do Rio (photo 14). This whole long process took place without obtaining authorisation to run the desired courses (Geography, History and Literature).

In 1993, FAESCI (photo 15) received a visit from the State Education Council (CEE) to check "in loco" the conditions for authorising and running the History, Geography and Languages courses. With a favourable opinion from the Council, dated 22 December 1993, the process of authorising and running the courses was forwarded by the State Department of Education to the Ministry of Education and Culture. By Presidential Decree of 4th March 1994, the History, Geography and Languages courses were authorised and, in the same year, the institution's first entrance exam took place, with classes starting in April 1994.

At the end of the second half of the 1990s, the Celso Inocêncio de Oliveira State College (FAESCI) was annexed to the UEG - State University of Goiás (created by State Law 13.456/1999). In 2000, the old name of FAESCI disappeared and was renamed the Pires do Rio University Unit of the UEG - State University of Goiás (photo 16).

According to Tavares and Marinho (2006, p. 58),

> The first graduates from the Geography, History and Languages courses graduated in 1997, but these three classes were only recognised in 2002 by EEC Resolution No. 251. For those graduating between 1998 and 2002, recognition came in 2000, by State Decrees No. 5.331 for the Geography course; No. 5.338 for the History course; and No. 5.383 for the Letters course. The Renewal of Recognition for the Geography, History and Languages courses for the 2003 and 2004 graduating classes was granted by means of EEC Order No. 1379 of 30 July 2004. Only the Geography course has already been granted Renewal of Recognition until 2006, with CEE Decree No. 2,103 of 1 December 2004; and the processes for Recognition of the History and Letters courses are still underway. In 2006, the Pedagogy course (established in 2000) received an on-site visit from the EEC Commission for its first recognition.

Photo 14: Bust of Celso Inocêncio de Oliveira Source: CARNEIRO, V. A. (2008)

Photo 15: FAESCI inauguration plaque-1993
Source: CARNEIRO, V. A. (2008)

Photo 16: UEG-Pires do Rio inauguration plaque / 2000
Source: CARNEIRO, V. A. (2008)

The same authors (2006, p. 130) go on to say:

> It was in the context of reflections on public policies for teacher training and the internalisation of higher education that the Celso Inocêncio de Oliveira State College (FAESCI) offered its degree courses in Geography, History and Languages in 1994 and, in 2000, as part of the process of creating the UEG, its Pedagogy course.

At UEG-Pires do Rio, 40 places/year are offered for the evening shift only.

The degree course in Geography must be completed by the student in a minimum of four and a maximum of seven years, totalling a workload of 3,224 hours (PEDAGOGICAL PROJECT OF THE GEOGRAPHY COURSE/UEG, 2005, p. 21). According to the Academic Guide (2006, p. 3) and the Academic-Pedagogical Guidelines (2007, p. 3) of the UEG- Pires do Rio, the Degree Course in Geography:

> It trains graduate geographers, with the main focus on teaching in primary and secondary

schools. The Geography teacher must actively participate in the organisation and management of the school, developing collective participation and decision-making skills, whether in the development of educational projects in the field of Teaching, Research and Extension in Geography, or in educational or environmental management, showing a predisposition to academic, cultural and scientific activities, to associate, approach, dissertate and synthesise problems linked to the relationship between society and nature. As well as teaching, Geography graduates will find a promising field of work in the various organisations that study and monitor the environment.

With regard to the Curriculum Matrix of the Degree Course in Geography, in the formerly FAESCI-Faculdade Estadual Celso Inocêncio de Oliveira, was authorised to operate in 1994, with the following matrix (chart 3):

Chart 3: Curriculum Matrix of the Geography Degree Course/FAESCI-1994.

CURRICULUM MATRIX FOR THE GEOGRAPHY DEGREE COURSE/FAESCI-1994			
Subjects -1st year	**Workload**	**Subjects - 2º year**	**Workload**
Portuguese: Writing Techniques	120	Physical Geography I	120
Organisation of Intellectual Work	60	Biogeography	60
Sociology	60	Human Geography I	120
Theory and Method in Geography	120	Regional Geography I	120
General Geology	60	Cartography II	120
Cartography I	120	Cultural Anthropology	60
Statistics	60		
Subjects - 3º year	**Workload**	**Subjects - 4º year**	**Workload**
Physical Geography II	120	Geography of Goiás	120
Human Geography II	120	Economic Geography	120
Regional Geography II	120	Special Geography Didactics	60
Geography of Brazil	60	Geography Teaching Practice	120
Educational Psychology: Adolescence and Learning	60	Initiation to Geographical Research and Field Practice	120
Didactics	60		
Structure and Functioning of Primary and Secondary Education	60		
Total Workload: 2,340 hours			

Source: Academic Secretariat of UEG-Pires do Rio, 2000
Organisation: CARNEIRO, V. A. (2008)

The 1994 degree course in Geography did not have a Political Pedagogical Project, as there were only simple academic reports contained in the appendix to the Internal Regulations of the Celso Inocêncio de Oliveira State College - FAESCI (1994, not paginated), establishing a degree course with a qualification in "Full Degree in Geography", with a minimum of four years and a maximum of seven years, with a total workload of 2,340 hours for academic graduation.340 hours for academic graduation, on a serial basis (annual subjects) and aimed at teaching in primary and secondary schools.

As was to be expected, the aforementioned Regiment (1994, not paginated) made no mention of field activities. In this case, it was up to the individual teacher, as well as the bureaucracy of transporting the students to the desired locations.

Looking through the course diaries, only in the diaries for General Geology, Physical Geography I and II, Biogeography and Initiation to Research and Field Practice, from the 1994 Geography Course Curriculum Matrix, were we able to find some notes about

visits to neighbourhoods in the city of Pires do Rio and neighbouring towns, rural areas and degraded watercourses (erosion, deforestation, silting, etc.) in the Railway Region and a trip to the Ecos Cave (Brasília surroundings).

The old Curriculum Matrix remained in place until 2003. A new Unified Curriculum Matrix was altered and implemented by UEG for the ten Geography Degree Courses at the University Units of Formosa, Goiás, Itapuranga, Iporá, Minaçu, Morrinhos, Pires do Rio, Porangatu, Quirinópolis and Anápolis in 2004.

In 2004, academic work began on the Unified Curriculum Matrix (table 5) for the Geography Degree Course (ORIENTAÇÕES DIDÁTICO- PEDAGÓGICAS, 2007, p. 4-6; PROJETO PEDAGÓGICO DO CURSO DE GEOGRAFIA/UEG, 2005, p. 21).

The Degree Course in Geography at the Pires do Rio University Unit of the State University of Goiás has a workload of 3,224 hours for students to complete (PEDAGOGICAL PROJECT OF THE GEOGRAPHY COURSE/UEG, 2005, p. 21-23), distributed as follows (table 4):

GENERAL DISTRIBUTION OF THE WORKLOAD OF THE GEOGRAPHY DEGREE COURSE	
Subjects	**Workload**
Specific	1.920
Pedagogical	976
Options	128
Complementary Activities	200
Total	**3.224**

Table 4: General distribution of the workload of the Geography Degree Course
Source: Political Pedagogical Project of the Geography Course/2005
Organisation: CARNEIRO, V. A. (2007)

For better clarification and detailing of the general distribution of the workload of the Degree Course in Geography, please see below:

1- Specific Subjects: made up of compulsory subjects that guarantee training for the graduate geographer, totalling 1,920 hours.

2- Pedagogical subjects: includes the pedagogical subjects plus Supervised Internship I and II, totalling 976 hours.

3- Optional subjects: at the beginning of each academic year, it will be the responsibility of the Course Coordinator in collegiate session to choose the optional subject(s) of interest to the University Unit with a workload of 128 hours; either a single subject of 128 hours or two subjects of 64 hours to be developed throughout the Degree Course in Geography.

UNIFIED CURRICULUM MATRIX FOR THE GEOGRAPHY DEGREE COURSE (UEG-2004)			
1st year		**2º year**	
Subjects	**Charge Hours**	**Subjects**	**Workload**

Education and Society	64	Public Policies in Education	128
Educational Psychology	128	Thematic Cartography	128
Systematic Cartography	128	Geomorphology	128
General Geology	128	Geography of Brazil	128
Theory of Knowledge and Geography	128	General Didactics	128
Study Guidance at Higher Level	64	Regional Theory and Regionalisation	128
Optional subject(s)	128	Complementary Activities II	50
Complementary Activities I	50		
3º year		**4º year**	
Subjects	**Charge Hours**	**Subjects**	**Workload**
Climatology	128	Political Geography and World Space	128
Biogeography	64	Cultural Geography and Social Movements	64
Agricultural Geography	128	Geography of Goiás	64
Geography and the Production of Economic Space	128	Urban Geography	128
Teaching Practice I	64	Teaching Practice II	64
Research Seminar I	64	Research Seminar II	64
Complementary Activities III	50	Complementary Activities IV	50
Supervised Internship I	200	Supervised Internship II	200

Optional subjects	**Charge Hours**
History of Geographical Thought	64
Methods. Techniques and Monitoring Applied to the Dynamics of Nature	64
Computer Science Applied to Geography	64
Text Production	64
Hydrography	64
Population Geography	64
Geography and the Environment	64
Literature and Geography	128
Hydrography Applied to Microbasin Studies	64
Statistics	64
Fieldwork in Physical Geography	64
Soils: Formation. Management and Conservation	64
Education Geography and Solidarity	64
Environmental Impacts in Cerrado Areas as a Result of Land Use Transformations	128
Introduction to Urban and Regional Development	64
Introduction to Geography and Tourism	64
The Quinary Period and its Environmental Effects	64

Chart 5: Unified Curriculum Matrix of the Geography Degree Course/UEG-2004
Source: Academic Secretariat of UEG-Pires do Rio, 2004
Organisation: CARNEIRO, V. A. (2008)

Complementary Activities: students must fulfil at least 200 hours of academic-scientific-cultural activities, which may include monitoring, extension courses, lectures, symposia, mini-courses, participation in scientific-cultural events, presentation of work at events, fieldwork linked to scientific and/or cultural events.

Students regularly enrolled on the Geography degree course at UEG-Pires do Rio must complete the course in 3,224 hours. The curricular matrix of the Degree in Geography offers subjects that are taught on an annual basis (1st, 2nd, 3rd and 4th years).

In the case of UEG-Pires do Rio, it should be noted that the optional subjects taken by the student in the first year of the course must total 128 hours, and from 2004 to the present day, two subjects have been offered, alternating between: History of Geographical Thought (64 hours), Geography and the Environment (64 hours), Education, Geography and Solidarity (64 hours) and Introduction to Geography and Tourism (64 hours).

Going back a little historically, it's worth remembering that on 4 December 2001, the 8th Plenary Session of UEG's Academic Council (CsA) regulated academic activities related to fieldwork in undergraduate courses, via Resolution No. 04/2001. Resolution No. 4/2001 of the UEG Academic Council (CsA), in Art. 1, emphasises: "the curricular completion of UEG undergraduate courses will also be achieved through fieldwork activities". Paragraph 1 emphasises: "Fieldwork is considered to be activities developed within a subject in the curriculum and aimed at empirically verifying theoretical aspects used in the academic training of undergraduates". Paragraph 2 emphasises: "academic activities will be carried out under the supervision of lecturers from UEG or another HEI, who have a command of the knowledge inherent in the field of knowledge to which the work refers".

Article 2 of this resolution defines: "Curricular academic activities will be considered to be fieldwork that is included in the course programme and that has been approved by peers and the Academic Council of the University Unit".

In the current curriculum matrix (table 5), we observed from the teachers' diaries that fieldwork is developed in each subject. We also found that in the current curriculum there is an optional subject called Fieldwork in Physical Geography.

We therefore believe that there is no need for a specific subject to deal with fieldwork. What we do need to be clear about is that fieldwork is a didactic resource that should permeate all the subjects in the Geography course and, if possible, in an integrated way.

In the next chapter, we will analyse the contributions, classifications and conceptions of the researchers, as well as the types of fieldwork that are implemented in Catalão and Pires do Rio according to the opinions of teachers and students.

CHAPTER 3

FIELDWORK: CONTRIBUTIONS AND CONCEPTS

The principle of this chapter is to show the various contributions, classifications and conceptions of researchers, students and teachers about the fieldwork resource. It will also summarise the results of the data obtained via a questionnaire from teachers and students at the State University of Goiás and the Federal University of Goiás.

3. 1 Humboldt's geographical acumen

Alexander Von Humboldt's scientific expedition from 1799 to 1804 to Spanish America was aimed at obtaining new in-depth knowledge about the "New World" for Europe.

The scientific objectives were very clear and the trip provided geographical reports integrating social, socio-economic, political and economic geography facts, based on empirical field research (KOHLHEPP, 2006, p. 260).

Kohlhepp's view shows that:

> His journey, as well as marking the transition from the first voyages of discovery to a new phase of expeditions focussed on clearly and scientifically defined problems, marked the convergence of a new vision of the "New World" to the European public. The main objectives of Humboldt's endeavour were not based on the conquests of the state or the exploitation of natural resources for the exploitation of a colony by the metropolis (2006, p. 261).

It goes on to say that:

> The explorers broke scientific paradigms with the help of their analytical procedures based on countless measurements and methods for quantifying observations, as well as the use of modern instruments combined with complex synopses in the form of topographical maps. These were more precise than any other known at the time. In addition to the maps, detailed profiles were drawn up of the landscapes they visited. The comprehensive collection of thousands of botanical specimens not only served to discover new species, such as barometric verification of altitude and temperature differences, but also to create a three-dimensional view of the differentiation of natural and cultural areas in high tropical mountain ranges (KOHLHEPP, 2006, p. 261). .

It emphasises that:

> Humboldt acquired intense knowledge of geology, mineralogy and some experience in the economic evaluation of mineral deposits in his rapid career as a mining specialist for the Prussian Civil Service and the valuable experience of his studies at the Mining Academy in Freiberg, Saxony (1791/1792), a school also attended by important geologists from the colonies of Spanish and Portuguese America (KOHLHEPP, 2006, p. 261-262).

So we can say that Humboldt travelled to Italian and Swiss soil in 1759,

> They already characterised scientific research trips, where he was able to put his existing geological, physical-geographical, phytogeographical and astronomical knowledge into practice through observation and contact with researchers in Geneva, as well as testing the latest instruments, acquiring practical experience in determining location and altitude, as well as drawing up soil profiles and maps. He was also concerned with these geographical studies when he was in Spain awaiting his departure for the New World (KOHLHEPP, 2006, p. 262).

The Humboldtian expedition took place before the political change in the country. Latin America. Humboldt witnessed the colonial, feudal and slavery periods and criticised them fiercely, as their social conditions were unbearable and inhumane. For this reason and for the scientific results obtained during his expedition to the tropics, as well as the innovative impulses he gave to geographical science and many other sciences, he is still admired for his erudition in Latin America and Europe today.

However, Alexander Von Humboldt (1769-1859) made a decisive impact on the In the first part of the 19th century, one of his greatest achievements during his only visit to the Americas was to observe and understand very clearly the relationship between man and the physical environment that surrounded him.

Long before he came to America, the young Humboldt was already showing some essays and

according to Azevedo (1959, p. 56),

> His first field research as a naturalist dates from this period: at the age of 21, he published a study of the basalts of the Rhine region (Mineralosgische Beobachtungen uber einige Basalte am Rhein, 1790); three years later, he studied the flora of the Freiberg region (Flora Fribergensis, accenunt Aphorismi ex Doctrina, Physiologia Chemica Plantaruum, 1793).

It also states that:

> He was no longer the sad captive bird of Tegel Castle. He began his professional life, at first as a mining inspector, then as a mining counsellor for Prussia. And he manages to travel outside his country's borders. Through the hands of Georg Forster, who had accompanied James Cook on his second voyage to the Pacific, Humboldt travelled to Belgium, crossed the English Channel and set foot on English soil. He then visited Austria, travelled through the Alps and entered Italy (AZEVEDO, 1956, p. 56).

In this context, we can highlight the great technical and scientific support of the doctor and naturalist Aimé Bonpland, who made a significant contribution to the legacy of the master Humboldt.

Azevedo emphasises that Humboldt, "joining the physician and botanist Aimé Bonpland ended up going to Spain, where, finding the support of the government, he definitively planned his great journey to Spanish America" (1956, p. 56).

Humboldt, along with the Frenchman Aimé Bonpland, travelled through Spanish America and didn't visit Brazil, but only a few stretches of the Amazon frontier, because the Portuguese wouldn't let him. The countries he visited were still under Spanish rule and production was based on slave labour. Humboldt made many scientific contributions to the most diverse fields of knowledge, but above all it was in the field of Geography that Humboldt's studies had the greatest impact and ended up influencing many other naturalists in their geographical explorations.

According to Kohlhepp,

> Alexander Von Humboldt knew that continuing his journey from the Orinoco River to the interior of the Amazon River systems in Brazil, claimed by Portugal, would be impossible. Portugal was guarding its colony - Brazil - against Spanish domination. As the Portuguese feared that Humboldt was a spy, the authorities in Rio de Janeiro, under orders from the King of Portugal, placed an order to imprison Humboldt if he entered Brazil. A few years later, when the Prince Regent, due to the occupation of Portugal by Napoleonic troops in 1808, took refuge in Rio de Janeiro, research opportunities could have arisen in Brazil for Humboldt. The expeditions of Maximilian, Prince of Wied-Neuwied (1815-1817) as well as the expeditions organised by Von Spix and Von Martius (1817- 1820) and many others in different regions of Brazil, after a short period of time, marked this scientific development (2006, p. 262).

For Azevedo.

> Unfortunately, we didn't have the privilege of receiving a visit from the great sage who, having reached the Casiquiare, could have naturally travelled down the Rio Negro and reached the heart of the Brazilian Amazon, had it not been for the measures taken by the Portuguese authorities. For this reason, other areas of tropical America benefited from his observations and detailed research (1956, p. 6061).

According to Monzón (2003, p. 45),

> Of all the travellers who have visited the Venezuelan territory over time, Humboldt is the one who has given us the broadest, most objective and systematic view of our

physical and human environment. As a result of his journey through various regions of America between 1799 and 1804, he produced a series of works written individually or with the collaboration of his travelling companion, Aimé Bonpland.

It also states:

> The importance of the study of travellers as a geohistorical source lies in the fact that: they study the landscape, space, society and its variations; they describe the sources of communication (roads, sites, villages and towns); they perceive the change in the cultural and natural space; the lack of studies of certain regions makes the testimonies of travellers the primary source of information about these places (MONZÓN, 2003, p. 44).

Due to medical problems on board, the ship stopped in Cumaná (Venezuela) in July 1799. This was not an obstacle,

> Since for Humboldt the "laboratory of nature" was available anywhere, the stopover turned into a 16-month stay in Venezuela, since an expedition to the interior could take several months. Humboldt explored the llanos and the Orinoco River Rainforest, travelling along the long uncertain connection between the Amazonian rivers, the fork of the Casiquiare River that leads to the Negro River (KOHLHEPP, 2006, p. 263-264).

According to Kohlhepp,

> Humboldt's expedition took place shortly after the political changes in Latin America that led to the end of the colonial period, except in Cuba during the first two decades of the 19th century. Humboldt would still see the global colonial economy based on slave labour. However, he strongly criticised this structure as a person who had internalised the ideas of the French Revolution. This is certainly why Humboldt's name is revered in many Ibero-American locations even today (2006, p. 263).

Humboldt's scientific contributions and versatile spirit made it possible for him to delve into various scientific fields:

> Countless measurements were carried out to determine location through precise barometric determination of air pressure above sea level, temperature, humidity, geomagnetism, air electricity, etc., and many other geophysical and meteorological values were determined. However, these revolutionary quantification methods and their interpretations, as well as his numerous topographical specifications, have led many non-geographers to mistakenly point to these facts as Humboldt's most important contributions to the field of geography. In fact, the central contribution of his works was the discovery of the "interaction of forces" (1808) in space that required individual assessments and precise analyses of these forces. Humboldt was not only a natural scientist and the founder of Physical Geography, but he should also be credited with the modern ecological approach, having explored the relationship between man and nature not in the sense of determinism, but from the perspective of correlations and co-operations from a synthetic point of view. "He was a genius at productive synthesis" (PLEWE, 1970 apud KOHLHEPP, 2006, p. 266).

We also noticed that from the

> Two "political essays" on New Spain (Mexico and the south-west of the United States) and Cuba were particularly important and form the basis of modern geography. In these works - in the best possible way - Humboldt presented modern, objectively orientated regional studies, the basis of Modern Geography, the first time with socio-geographical content in a systematic way. Thanks to his precise observations in Mexico, these principles could be put into practice, for example: in investments in mining and roads. They were used by the Viceroyalties of New Spain as a basis for political and economic decisions. Humboldt exhausted existing statistical sources to publish the most important data and believed there was nothing better in Europe at the time (SCHMIEDER, 1964; STEVENS-MIDDLETON, 1956 apud KOHLHEPP, 2006, p. 267).

Humboldt, in his legacy,

> As a physical geographer, founder of climatic and plant geography (phytogeography), as an author of regional studies with a strong commitment to the relevant aspects of human and geopolitical geography, as a cartographer and with the very didactic graphic presentation of profiles, he set the standards for the modern development of geography as a science (KOHLHEPP, 2006, p. 272).

According to Hettner (1927 apud Kohlhepp, 2006, p. 272),

> A new era for geography dawned in 1799. This was the year that Alexander Von Humboldt began his great American journey. The greatest advances in Geography can be attributed to this journey: the foundation of various branches of General Geography as well as the foundation of scientific Regional Geography. This is to the everlasting credit of Alexander Von Humboldt who, on his great journey through South and Central America, recorded and presented the full character of the countries and their inhabitants in a splendid way.

Beck (1996 apud Kohlhepp, 2006, p. 272) praised him as "the greatest geographer of modern times". Because of his career and geographical knowledge, "Carl Ritter considered Humboldt to be the scientific discoverer of America" (KOHLHEPP, 2006, p. 272). Finally,

> It would be fair to call Alexander Von Humboldt, with his holistic understanding of nature and the scientific method, the founder of research into the tropics based on scientific methods. Research that went beyond the mere description of "exotic" details and took a step towards the first approximations of a systematic tropical ecology. Today, research on the tropics and Latin America in Germany is carried out in the light of this tradition (Meyer-Abich, 1993 apud KOHLHEPP, 2006, p. 273).

Corroborating Azevedo, we note that "for all these reasons, as De Martonne rightly observed, Alexander Von Humboldt cannot be considered just a naturalist and a traveller, but a true geographer" (1956, p. 71). In line with Helferich,

> The fact is that Humboldt helped create the world as we know it, and his influence is felt all over the planet, even where his name is not often remembered. The product of a rich cultural tradition that has its origins in the ancient Greeks and encompasses such disparate titans of the Enlightenment as Francis Bacon, Isaac Newton, René Descartes and Immanuel Kant, Humboldt passed this tradition on to his own successors in science, including Charles Darwin, Albert Einstein, Max Planck and Edwin Hubble. Humboldt's physique génerale is a link in the conceptual chain that comprises fundamental theories such as evolution, relativity, quantum mechanics and the big bang (2005, p. 24).

He goes on to say,

> Even in the midst of today's rampant scientific specialisation, Humboldt's quest for the "unity of nature" not only survives but thrives, still producing some of the most exciting areas of contemporary research, such as cosmic strings, complexity, emergence and the elusive "theory of everything" (HELFERICH, 2005, p. 24).

According to Alves (1997, p. 85),

> Fieldwork has been an important tool in the geographer's study. If we take a look back at the history of geographical thought, we will notice that the practice of observation has become an essential resource for geographers of all generations and nationalities. Just to remind us, Geography was born as a science, valuing this spirit of seeking elements in direct observation to better understand the landscape. Humboldt, for example, was one of those geographers who first used this tool to explain the phenomena that manifested themselves on the surface of the earth.

He goes on to say:

> In fact, Humboldt's vast body of work was largely based on the observations he gathered during his travels around the world. The most interesting thing about this author's work is that, despite being based on observations, it is not characterised as empiricist, as Humboldt was able to combine empirical and abstract work (ALVES, 1997, p. 85).

It also highlights:

> Another important element in Humboldt is the fact that he considered that observing the landscape should not be a cold, emotionless exercise, but just the opposite, that is, nature should be observed with feelings, that is, contemplated in the most pleasurable way possible, and for this to happen the observer had to let all their sensitivity flow (ALVES, 1997, p. 85).

He also considers that:

> The legacy left by Humboldt with regard to the practice of fieldwork, in my opinion, is an important lesson for us geographers, especially for those who adopt this procedure as an additional resource for unravelling the landscape. In this sense, I believe that fieldwork can indicate ways of obtaining new interpretations of the reality being studied. However, it is important to be aware of the fact that the act of observing, for the geographer, should not take place without emotion and even less without a commitment to the reality being investigated; at the same time, it is necessary to put into practice the exercise of observing, combined with that of feeling and reflecting (ALVES, 1997, p. 85).

In this way, I can see that Humboldt's contributions, methodological and scientific characteristics, via going into the field, are part of the geographical struggle. We can't fail to recognise the Humboldtian approach within geography, because if we take the opposite view, it would lead us to believe that certain professionals suffer from "geographical glaucoma".

The types of fieldwork will be presented below, as well as their conceptions, characteristics and classifications according to their authors.

3.2 Types of Fieldwork

Throughout the history of Geography, as well as other disciplines, fieldwork has been frequent and undeniably considered an important way of improving and perfecting knowledge.

Professor Santos (1999, p. 112) makes the following comment:

> We can say that geographers at the beginning of the 20th century had important empirical experiences when analysing the realities they were researching. In the field, they were able to access information that at times seemed to repeat itself in a boring way, but in other realities turned out to be different, new and, in many ways, surprisingly fascinating.

Coltrinari says that "Sauer and Cholley are pillars that support the recognition of the role of fieldwork as an instrument for verifying and recording changes in landscapes and as training for the eyes and minds of future geographers" (1996, not paginated).

He also reports:

> But it's not just about collecting with your eyes, hands or instruments during the empirical phase of building knowledge. Without theories or hypotheses in mind, fieldwork, whether it's a reconnaissance excursion, a periodic interview campaign or process monitoring, runs the risk of being boring and tiresome and, quite rightly, criticised or rejected (COLTRINARI, 1996, not paginated).

Sauer (1956) and Cholley (1942), quoted by Corrêa (1996, not paginated), emphasised "the importance of fieldwork in the training of the geographer, which trains the eye to see and the mind to generalise".

Cholley (1942 apud Corrêa, 1996, unpaginated) states that "nothing is more valuable for the training of the geographer than contact with reality through, in part, fieldwork, both for pedagogical purposes and for research".

He also stresses that "in reality, fieldwork is a tradition whose importance is recognised by everyone, especially those who see the natural or cultural landscape as the objectification of geography" (CORRÊA, 1996, not paginated).

Santos says,

> What we want to discuss is what we can put into practice in fieldwork. When we want to approach a particular problem scientifically, we draw up our theoretical-methodological framework. This framework is often considered to be a scientific guide that will guide us through rigour without fail. But there are many ways of analysing a given reality, and it's important that we don't recognise just one method. It is clear that we will have to resist the academic indoctrinations that still insist on adopting a single method (1999, p. 118).

For Cruz, "a phrase from Prof Yázigi (USP), repeatedly passed on to his students, clearly reveals the importance of field research in Geography: NOTHING REPLACES FIELD WORK" (1997, p. 93).

The purpose here is to get to know the methodology used during fieldwork by the various authors working in this area. Thus, we come across different classifications of fieldwork. Depending on how each teacher plans their fieldwork, the activities can be very rigorous or left to chance.

In short, "fieldwork, although carried out in the same region, can therefore follow different plans" (STERNBERG, 1946, p. 14-15). However, some authors in geography and geology have grouped fieldwork into categories.

It is worth mentioning and making it clear that fieldwork will be considered synonymous with the expressions: field activity, field lesson, field trip, field outing, incursion, trip and geographical excursion.

Firstly, we have Carneiro and Campanha (1979 apud Compiani and Carneiro, 1993, p.

90), proposing the following classification:

> **Illustrative**, whose central aim is to illustrate the various concepts seen in the classroom; **motivational**, where the aim is to motivate the student to study a certain topic; **training**, which aims to guide the execution of a technical skill; and **problem-generating**, which aims to guide the student to solve or propose a problem (my emphasis).

Clarifying a little further, Costa (2006, p. 18-19) states:

> **Illustrative activity**, when the various concepts, seen in the classroom beforehand, are only illustrated to the students in the middle;
> **Motivational activity**, when the aim is to motivate the student for a particular topic that will be covered later in the lesson;
> **Coaching activity**, when the teacher wants to guide the performance
> **Problem-generating activity**, when the student is guided to solve or propose a problem (my emphasis).

Fernandes et al. (1981 apud Pereira et al., 2003, not paginated),

> They add a type of field, which they call the roles of fieldwork, which is **problem-generating**. This consists of getting the student to create a problem, which involves analysing the integral parts and their interrelationship. This fieldwork role is understood as a complex approach, although in some cases it may be less comprehensive (my emphasis).

In Brusi (1992 apud Compiani and Carneiro, 1993, p. 91), there is a classification into three groups, according to the roles assigned to the teacher and students during fieldwork: directed excursions, semi-directed excursions and self-directed excursions.

Costa (2006, p. 20) analyses Brusi's proposal (1992 apud Compiani e Carneiro, 1993, p. 91), details that the:

> **Directed excursions**, where the teacher is a lecturer, "says it all", the students just listen to the teacher and take a passive role during the activity, the teacher is the protagonist;
> **Semi-guided excursions**, with the "Socratic" teacher or using teaching guides, the teacher uses other elements to enrich the visit from a scientific point of view, as he is supported by someone who knows the place;
> **Self-directed excursions**, when students lead them for themselves, are student-centred, the student plays an active role in the visit and consequently in their learning during the activity (my emphasis).

According to Compiani and Carneiro (1993, p. 94-96), geological and/or geographical excursions are classified according to their didactic role:

> **Illustrative**: it is traditional and reaffirms knowledge as a finished product; **inductive**: it aims to sequentially guide the processes of observation and interpretation so that students can solve a given problem; **motivational**: it aims to arouse students' interest in a given problem or aspect to be studied, valuing the experience of each student and their questions; **training**: aimed at training skills, usually with the use of apparatus, instruments or scientific apparatus; **investigative**: this type of field trip allows students to solve certain problems in the field and the teacher fulfils the role of guide (emphasis mine).

Scortegagna (2001, p. 25-31) supports the proposal of professors Compiani and Carneiro (1993, p. 94-96), but centres his ideas on the geology/geography interface, presenting two types of fieldwork: the "generic field trip" and the "autonomous field trip".

According to the same author (2001, p. 28-31), there are two field trips:

> **Generic field trip**: this type of activity is common in Geography courses and refers to excursions where the main objective is to get to know a particular region, possibly not yet visited by the majority of students and teachers, and is generally not linked to a subject, but to the course and is most often carried out at the end of the school year; **Autonomous field trip**: this outing aims to awaken the student's investigative spirit and prepare them for their future professional reality, it is usually carried out, preferably, in the region where the students are, in areas chosen by them and without the presence of the teacher; the investigation is constant, with the teacher playing the role of advisor (my emphasis).

Suertegaray (2002, p. 104-106),

> He warns that fieldwork is not everything, as it suffers from limitations that will be overcome by other forms of apprehending knowledge. His classification covers the fieldwork usually carried out in Physical Geography: a) **generic reconnaissance of the place or places (excursions)**: these generally have a general character of reconnaissance, description and observation training; b) **specific reconnaissance of elements or phenomena in the field (field exposure)**: these are done from a script, where the teacher, previously establishes the places to be observed (observation points); at these points, in general, an exposition is made about what is observed, from the teacher who guides the

> work; c) **recognition of the place** from the selection, a priori, of procedures that imply information gathering by the group involved (field survey) and d) **recognition**, in the field, **of patterns observed in images of places** (photographs and/or images) (my emphasis).

For Carvalho, the geography excursion is aimed at "contact with reality, which in itself determines the beginning of an entire learning process" and "a good excursion, well executed, is equivalent, in my opinion, to many lessons" (1941, p. 98) (my emphasis). He went on to point out that "if each unit of work in Geography could be preceded by an excursion, appropriate to the subject in question, and concluded with another excursion to fix and revise, I am sure that the Geography Course would be a success" (CARVALHO, 1941, p. 98).

For Ruellan, "research in the field makes it necessary to organise excursions of two very different types: the reconnaissance excursion and the detailed investigation" (1944, p. 36). It should be noted that in Ruellan (1944, p. 36),

> The **reconnaissance excursion**, which can last from a few days to a few weeks, is designed to reveal to the researcher the essential features of the region they are trying to study. Its originality comes from the fact that it has a defined itinerary, travelled in a short time and during which all the problems arise together, which requires a great deal of experience. This is why beginners must be guided by an experienced researcher at the same time.
>
> The itinerary adopted should, of course, allow you to cross the region by cutting through the essential landforms, so that you can grasp the contrasting aspects and transitions it presents from both a physical and human point of view. It is also necessary to use the summits during this itinerary, which offer extensive views, and to look for opportunities to observe the subsoil: valleys; road cuts, etc (emphasis added).

For the reconnaissance excursion, it is also important to recognise and analyse the elementary forms of the relief; the need for a topographical survey; the need to research the relationship between the relief and the geological structure; climatological research; hydrographical research; biogeographical research and research into human geography (RUELLAN, 1944, p. 36-40).

According to Ruellan (1944 apud Sansolo, 1996, unpaginated), detailed investigative fieldwork aims to "carefully control the observations and interpretations of predecessors, to confirm them, complete them, graduate them or show what is inaccurate about them, in order to arrive at the realisation of an original overall work for all that it brings to the subject" (my emphasis).

Sansolo further emphasises:

> For detailed fieldwork or data collection in geographic research, prior knowledge through various sources of information, whether bibliographical or the results of previous research, whether aerial photos or ground-level photos, maps of various scales and even reconnaissance fieldwork, is a procedure that can avoid various problems, especially those of a logistical nature, as well as facilitating and elucidating technical and theoretical issues (1996, not paginated).

In his research work, Sansolo (1996, not paginated) uses the classifications of both Ruellan (1944) and Carvalho (1941) on fieldwork in Geography.

Regarding fieldwork methods, Sternberg (1946, p. 14-15) says:

> There are countless types and variations of geographical fieldwork, as there are many conditions that regulate their nature.
>
> Thus, for example, fieldwork varies according to the level at which it is carried out: we can have **pure research geographic work** or **just didactic geographic work**; a combination of these two modalities is also common.
> Geographical research excursions, on the other hand, can be **utilitarian in nature** and have economic objectives in mind (studying a region for this or that type of exploitation, for example), or **speculative in nature** and have purely scientific objectives.
> In the case of excursions for teaching purposes, the work will be quite different depending on whether it is carried out with beginners or with students.
> Fieldwork will also vary according to the point of view adopted: if an **analytical attitude** is adopted, greater development can be given to one of the branches of Geography (Physical or Economic Geography, for example); if the study is **synthetic**, it will be carried out along the lines of Regional Geography (my emphasis).

Silva (1982, p. 49) presents three types of fieldwork in Human Geography: "empirical analytical work, work with a logical approach and epistemological and ontological dialectical analysis". In Silva, the method of empirical analytical work:

> It is made according to experience, which is the basic parameter of judgement.
> This research must be supported by oral or written knowledge, which can include testimonies, cartographic documentation, aerial photos, profiles, etc.
> Fieldwork then involves observing the landscape and collecting data, depending on the researcher's objectives. For this reason, excursions and direct contact with the population are important, and questionnaires or interviews can be carried out.
> According to supporters of this type of fieldwork, true knowledge can only come about through direct contact, which includes "feeling" the situation being researched (1982, p. 50) (my emphasis).

Next up:

> The method of **work with a logical approach** in Human Geography is a result of the construction, mathematical or otherwise, of models, which represent an "ideal" reality and operate as hypotheses and theories. In their construction, these models receive the treatment of the scientific method - hypothesis, observation, analysis and generalisation - which allows the analysis of the reality represented from this resource (CORRÊA, 1982, p. 50) (emphasis mine).

There are also works with epistemological and ontological dialectical analysis, which have a different conception from the previous ones:

> His assumption is that theoretical knowledge of reality is possible from the apprehension of the basic categories of being.
> On the epistemological side, as in Bachelard, there is the consequence of a genetic structuralism, which indicates the categories in an objective idealist sense. In this case, the field is a concrete that is very close to that of logical conceptions, although the intention is to work with the essence of reality. In some cases, a systemic structuralism emerges, as in Milton Santos (POR UMA GEOGRAFIA NOVA).
> Dialectical ontological analyses have two variants: the orthodox (hermeneutic) and the modern (analytical). The first corresponds to the idea of affirming the essence of a being, as in Marx, for example, when he says that categories are manifestations of being, determinations of existence. The second is presented as a dialectical, analytical construction, in which ontology, as an affirmation of being, is complemented by affirmations about its appearance; in discourse, then, there are affirmations of the type of Marx's, such as the one enunciated, and also definitions of a logical nature.
> The concept of field in dialectical epistemological analysis, as in Milton Santos (POR UMA GEOGRAFIA NOVA), appears as an essence that is immediately (phenomenologically) perceived. Santos' analysis, although immanent, refers, as he says, to what is happening in front of our eyes.
> The concept of the field in the orthodox variant of ontological dialectical analysis immediately appears in the form of manifestations of reality, manifestations that are a form whose content is not directly given. It is a question of grasping the movement of

reality through an analysis of necessary and determined relationships, of which people are not necessarily aware and which are independent of this awareness. In this case, it is existence that determines consciousness and not the other way round, as in Marx.
Fieldwork in epistemological dialectical analysis consists of the direct or indirect collection of information, which is organised structurally, according to sequences articulated by their apparent logic, referring to the essence of reality.
Fieldwork in orthodox dialectical analysis consists of ontological reflection on the totality, the apprehension of the elementary categories of the phenomenon under study, which already contains the whole. Hegel's "simple concept" is, from the outset, a totality.
Fieldwork in (modern) analytical dialectical analysis separates the two moments of the orthodox method. Therefore, there is a movement of reflection that accompanies a moment of collecting empirical information, which is taken as manifestations of reality. Therefore, here too, as in orthodox analysis, knowledge is always an approximation (CORRÊA, 1982, p. 51).

Corrêa points out that "fieldwork then becomes a problem of many fieldworks" (1982, p. 52). He argues that "for this to happen, Human Geography must carry out the fieldwork necessary to verify its chances of success. And it can count on the fact that the crisis it is experiencing brings with it the elements that lead to a solution" (CORRÊA, 1982, p. 53). He goes on to say that "fieldwork in Human Geography has become a complex issue at a time when society has also become complex" (CORRÊA, 1982, p. 53). He also points out that "teaching must be democratised, without access to fieldwork itself becoming impossible" (CORRÊA, 1982, p. 53).

According to Suertegaray (1996 apud Amorim, 2006, p. 20), the following are presented

different types of fieldwork:

Excursions: consist of a general reconnaissance of the place or places. In general, they have a general character of recognition, description and observation training. Although they are useful from a didactic point of view, they can be criticised for their superficiality.
Exposures in the field: these consist of the specific recognition of elements and phenomena in the field. These are carried out using a script in which the teacher establishes observation points in advance. At these points, a presentation is made about what has been observed, based on what the teacher considers to be important for the work. Criticism of this modality is that the students only take on the role of observers and do not gain an understanding of the object of study. **Field survey**: recognising the place by selecting "a priori" procedures that involve the group involved gathering information. The advantage of this method is that the group can handle equipment, discuss data and ideas, and conclude on what has been observed. It also allows the group to get involved in the pursuit of an objective.
Tests: recognising in the field patterns observed in images of places such as aerial photographs and/or images. Its validity lies in the technical interpretation of these images, although it cannot be practised in isolation and must be associated with other forms of work (my emphasis).

Lacoste says,

The actual research work, observation in the field, is on a large scale and, at this level, it is only part of the phenomena that can be properly learnt; the others must be anticipated on a smaller scale and, for this, it is necessary to use representations that research in the field cannot provide. If fieldwork is not to be just empiricism, it must be linked to theoretical training, which is also indispensable. Knowing how to think about space doesn't just mean placing problems in a local context; it also means effectively linking them to phenomena that are developing over much wider areas. It is no less true that research, insofar as it corresponds to the extraction of an abstract from a concrete, through research and field observation, gives great importance to the level of conceptualisation on a large scale (of course, research can also start, above all, from already elaborated abstractions; the training of researchers is then different and much less hesitant) (1985, p. 20).

Therefore, it must be made clear that going into the field requires planning, rules and coherent pedagogical action in order to interpret realities. Care must also be taken so that it doesn't degenerate into mere sightseeing and unreasonable visits to parks, "zoos and safaris" (KAYSER, 1985, p. 35).

Another type of fieldwork widely used in urban areas, according to Gelpi and Schaffer (1999, p. 117), is the "urban route guide" (my emphasis). This activity,

> It is a unique opportunity to develop systematic observation, guided and explored by the qualified intervention of the teacher; and description, seen as the intellectual capacity to select, order and organise information, to indicate what is visible and to establish inferences, depending on the student's stage of development. Through the recording and representation of what has been observed, it allows creativity to be exercised, whether in verbalisation or textual production, or in the creation of drawings, sketches, models or the assembly of photos and figures. It opens up spaces for the student or group to establish opinions and make a critical appraisal, to problematise observed phenomena, to establish hypotheses and to move on to research. It is therefore a resource with multiple possibilities of approach (thematic and interdisciplinary) and multiple perspectives of exploration in the learning process (GELPI and SCHAFFER, 1999, p. 117-118).

According to Cavalcanti,

> The **environmental study** is a type of school activity that can be linked to a broad research activity, when it is one of its stages, or it can be developed as a specific procedure for dealing with geography content (2002, p. 90) (my emphasis).

He goes on to say that "the aim of environmental studies in teaching is to mobilise the students' sensations and perceptions in the knowledge process in order to

then proceed to conceptual development" (CAVALCANTI, 2002, p. 91). However, we can say that for the study of the environment to gain didactic value, it needs to fulfil a number of stages: "preparation; carrying out the work; exploring the work carried out in the classroom" (CAVALCANTI, 2002, p. 92).

This procedure is suitable for studying geography at school because of its didactic-pedagogical character, which they emphasise:

> a) favour geographical conceptualisation; b) enable the development of procedural skills related to measuring distances, heights, frequencies, etc.; c) develop observation skills; d) enable the development of interconnected views of aspects that are conventionally treated separately in teaching; f) they enable comparison, identification of similarities and differences between areas and landscapes; g) they enable an environmental perspective on the surroundings and a search for solutions to environmental problems; h) they constitute a unique framework for the development of cartographic skills (OGALLAR, 1995 apud CAVALCANTI, 2002, p. 92). 92).

With regard to the study of the environment, "the understanding is that this is just one stage in understanding geographical space" (CAVALCANTI, 2002, p. 91).

The researchers Pedrinaci et al. (1994) and Del Carmen and Pedrinaci (1997) cited by Costa (2006, p. 23-24), classified fieldwork into four types: "traditional field class, where the teacher is a cicerone; field class as autonomous student discovery; field class as teacher-led observation; problem-solving orientated field class".

Professor Costa (2006, p. 24-25) further outlines that:

> In the **traditional field lesson**, the student is seen as a blank page, the teacher is the

protagonist, whose concern is to fulfil the established plan and make an orderly transmission of knowledge, with the aim of reaching the student as directly and quickly as possible.

In the **autonomous discovery field lesson**, the student takes centre stage in the activity, and procedures, values and attitudes are now important.

In the **directed observation field lesson**, the teacher meticulously plans the outing, selects the places where each stop will be made, what kind of observations should be made in each place and how they should be recorded; with all this, he also carefully draws up a script that he will give to the students; although all the leading role up to this point has belonged to the teacher, during the outing this leading role will become the student's; the student will be the tutor, will be responsible for carrying out the established plan, will clarify any doubts about the script given to the students and will eventually help draw up some answers;

The problem-solving **orientated field lesson** was created to overcome some of the difficulties and limitations of the previous models; this type of fieldwork encompasses three stages with different characteristics: **pre-exit**, **exit** and **post-exit** (my emphasis).

In the pre-field trip,

> The activities began with the formulation of a problem. This is understood as a question that the student cannot solve mechanically by applying an algorithm, but which requires conceptual or empirical investigation. It is important that the problem formulated has a clear meaning for the student, that it is related to the content being worked on in class, that it allows relevant aspects of the curriculum to be addressed and that it can be approached a priori from one or more theoretical perspectives. Once the intent and meaning of the problem have been discussed and understood, the students should draw up a script that will constitute their observation hypothesis. At this stage, it is advisable to alternate the small group activity with a more general exchange in which the teacher demands concreteness, helps to define the contrast procedures that will be used, asks questions that require further clarification or the consideration of other alternatives. It's not necessary, or even convenient, for all the students to have the same observation hypothesis or contrast script at the end. The aim of the general exchange is not to unify the proposals, but rather to enrich them, clarify them, compare them and seek their internal coherence. Preparing in this way makes it easier to relate the ideas that each student has, favours motivation and initial discussion, avoids overly abstract planning and makes it easier to incorporate new questions or elements to observe based on the students' suggestions (COSTA, 2006, p. 25).

Leaving the field,

> This is the phase of comparing the hypotheses developed in the pre-exit. Here, each student will carry out the observations, measurements and notes they had defined. New problems may arise, some of which can and should be tackled on the spot, while others will have to be left with open questions and noted down so that they can be worked on another time. Each group of students has its own work plan and can work quite autonomously. The teacher will demand that the plan is followed or that changes to the plan are justified, will ask for objectivity and rigour in the observations that are made, will stimulate reflection, encourage reasoning and the justification of statements that are made, raise new questions, suggest other options and show some observations that may go unnoticed (COSTA, 2006, p. 26).

After leaving the field, the work is based on:

> Reflection on the whole process from start to finish. Students should write down the knowledge they acquire or modify, check their conclusions and communicate them to the class. The collective presentation and discussion of the results of the research carried out provides very important elements, such as the use of techniques and resources for expression and communication, the confrontation of ideas with the rest of the class and the enrichment that is not produced individually, but with a collective effort. The teacher's intervention at this stage should be to demand rigour in the conclusions, to facilitate exchange and comparison and, above all, to establish generalisations and indicate relationships with other content being worked on (COSTA, 2006, p. 26).

Similarly to the above, Torre (1991 apud Costa, 2006, p. 21) believes that the ideal would be to carry out "occasional, short and frequent outings, in which only one theme is analysed". He also thinks that the occasion requires "bringing together the various activities in a day, in which several observations would be concentrated", highlighting three moments for carrying out the field activity: "before the outing, the outing and after the outing" (my emphasis).

Orion and Hofstein (1994 apud Moreira, 2005, p. 2), in the same line of reasoning and characteristics about fieldwork, similarly to Torre's proposal (1991 apud Costa, 2006, p. 21), consider the following terminologies: pre-trip, trip and post-trip.

Following this path, Rodrigues and Otaviano (2001, p. 37) emphasise that when planning fieldwork, three fundamental and essential moments are considered: "preparation, implementation and results/evaluation".

Tomita (1999, p. 14-15) states that when planning fieldwork, the essential points involving the teacher and students should be emphasised.

> It's up to the teacher to ask questions:
> Where to go? What is your prior knowledge of the area? What are the proposed objectives? How to get there? What is the geographical content? Have you mastered the content? Have you mastered this working technique? Have you planned in detail? Are the students sufficiently prepared for this activity? What attitudes (way of thinking and acting) do you expect from the students? How do you assess whether learning has taken place during the activity and at the end?
> It's up to the students to prepare (with the teacher):
> What is fieldwork? What is it for? What is it for? Where to go? Why choose it? How to go? When to do it? What to take? What equipment? How to dress? Which committees are needed and what are their roles? What are the stages of the work? What results are expected and obtained? What are the proposals for future work?

It also clarifies that:

> During fieldwork, the teacher must remain a motivating link and arouse the students' interest by discussing and asking questions that pique their curiosity, in such a way that they feel the importance and necessity of this activity as a complement to the theoretical lesson (TOMITA, 1999, p. 14).

According to Cruz (1997, p. 93-94), prior planning of fieldwork is a prerequisite for the success of this activity. This planning must include: "defining objectives; preparing a plan/route and drawing up a timetable".

In Sternberg (1946, p.17),

> Geographical fieldwork is more complex than it might seem at first glance. In order to carry it out efficiently, careful preparations are needed. During the course of the work, a series of rules must be followed and it will be necessary to work with a set of techniques that must be mastered. Once the fieldwork itself has been completed, there is still the preparation and presentation of the data collected. In summary, fieldwork is divided into three successive and complementary stages: (1) planning and organisation, (2) implementation and (3) drawing up the results.

It is worth noting that, according to Castrogiovanni and Schutz, we can separate three stages to diagram the methodological behaviour of the teacher and students involved in fieldwork: "the before, the during and the after" (1986, p 46-47).

Planning is an important step in achieving success on a geographical excursion. According to Carvalho (1941, p. 98-100), there are five stages to follow: preliminary preparation, psychological preparation, organisation of the excursion, directed observation and reports.

In the case of preliminary preparation, it is necessary to outline the purpose, choose the points of visit and also "hence the need for the teacher to prepare the excursion beforehand", in order to avoid problems on the spot (CARVALHO, 1941, p. 98-99).

With regard to psychological preparation, the teacher must prepare them for an excursion of a geographical nature, where there are goals to be met, and it is not just a walk, much less "a recreational activity" (CARVALHO, 1941, p. 99).

The organisational part of the excursion falls to the teacher to work on the bureaucratic side of things (establishing rules, making contacts, drafting letters and authorisations, etc.) and to the students to collaborate, in a disciplinary and technical way, to ensure the activity runs smoothly (CARVALHO, 1941, p. 99-100).

When it comes to directed observation, the teacher will have to be skilful in training and showing the pupils an awareness of space, equipping them with the "faculty of seeing and observing", of "getting into the topographical environment", of "interpreting geographical landscapes" and punctuating the normal things of study. On this occasion, "directed observation is nothing more, as far as we are concerned, than a process of visual utilisation of the geographical environment for education" (CARVALHO, 1941, p. 100).

Finally, there are the students' reports, which are a document full of geographical insights for use in the classroom. Thus, according to Carvalho (1941, p. 100),

> In order to keep the hikers' attention during the tour, so that they are stimulated in their work of seeing, observing, noticing and counting, it is essential to clearly establish the obligation to present a report of what was done and recorded, under the teacher's supervision or suggestions.

According to Rodrigues and Otaviano, "contact with reality will give students a new dimension to the subjects covered in class which, if well planned and orientated, will serve, among many other purposes, to stimulate study in conjunction with the different subjects" (2001, p. 35). So,

> Therefore, fieldwork as a teaching resource is of primary importance, because it offers formative potential that should be taken into account in the teaching-learning process as one of the most accessible and effective pedagogical techniques for teachers (RODRIGUES and OTAVIANO, 2001, p. 36).

In Sansolo, we find that:

> Fieldwork generally requires some planning considerations: prior knowledge of the location; prior collection of secondary information; technical preparation of participants; drawing up a work plan; selection and preparation of field material. (1996, not paginated)

Researchers Marandola Jr. and Lima (2003, p. 178-179) note that:

> The morphological character of the landscape is didactic, allowing students, in **guided fieldwork**, to observe it while seeking, in their observation and analysis, the component elements of the space itself, which materialise their impressions of the landscape. It is in this sense that the landscape can enable a holistic view of space, allowing it to be observed and analysed in an integrated way (my emphasis).

They claim that:

> It is because of this holistic pedagogical approach that we point to the use of landscape as an enabling category for these processes, due to its morphological aspect. Using this concept in fieldwork and in Geography teaching in general will enable better conditions for knowledge to be developed in an integrated and non-disciplinary way, with a view to understanding the complexity that is inherent to space (MARANDOLA JR.; LIMA, 2003, p. 179).

Junker (1971 apud Marandola Jr. and Lima, 2003, p. 175) points out the stages for excellent fieldwork: "observation, recording, analysis and reporting".

This angle of vision on fieldwork is the same for anthropologists when they explore the labyrinths of native communities, both rural and urban.

According to Scortegagna, "fieldwork is of fundamental importance when learning geology and geography. It is in the field that students will be able to perceive and apprehend the various aspects surrounding their study, both natural and social" (2005, p. 37).

Morcillo et al. (1998 apud Scortegagna, 2005, p. 37), "believe that fieldwork is essential in the teaching of natural sciences, something apparently impossible to fulfil with classroom and laboratory activities".

From this point of view, "fieldwork in Geography and Geology is undoubtedly essential, as it allows students to position themselves in relation to theoretical knowledge and current reality, demystifying science and building knowledge that is closer to their everyday lives" (SCORTEGAGNA, 2005, p. 37).

Paschoale (1984 apud Scortegagna, 2005, p. 37) describes fieldwork as

"a scenario for generating, problematising and criticising knowledge, where the conflict between reality and ideas takes place with all its intensity".

According to Scortegagna, "particularly in Geography, field practices present infinite possibilities for research and investigation, since it is in geographical science that physical and human aspects become concomitant objects of study" (2005, p. 37). As for Souza (2003, p. 1),

> Fieldwork is an essential tool for geographers. It enables geographers to carry out their work better, because it allows them to see the phenomena they want to work with. What's more, it's extremely important for geography undergraduates to have contact with this work tool, as it allows them to visualise the theories they've seen in the classroom.

He also emphasises: "the idea of integrated fieldwork is a vision that seeks an alternative so that student learning is not jeopardised, since public universities are in crisis and the availability of resources is increasingly difficult" (SOUZA, 2003, p. 1) (my emphasis). It emphasises that:

> In this way, the aim of the work is to present a model script for fieldwork that is multidisciplinary, that can cater for disciplines in the area of Physical Geography and that is economically viable, in an attempt to circumvent the crisis in free public higher education" (SOUZA, 2003, p. 1).

Professor Sansolo emphasises:

> In order for fieldwork to be carried out and obtain satisfactory results, it is necessary not only to reflect on the theory of its importance for research, but also to plan carefully, taking into account an itinerary that makes it possible to observe contrasts in the landscape (2000, p. 141).

In Suertegaray,

> We therefore see fieldwork in a broader sense, as an instrument of geographical analysis that allows the object to be recognised and which, as part of a research method, allows the researcher to be inserted into the movement of society as a whole. This view does not deny the possibility of using instruments in the field and in research in general (2002, 95-96).

From Corrêa's perspective,

> Fieldwork must not become a trap for the geographer in the face of increasingly complex and elusive landscapes and spatial relations. It must now be more critical and theoretically grounded, as it was in the past, one of the main means by which the geographer learns to see, analyse and reflect on the endless movement of man's transformation in its spatial dimension (1996, not paginated).

Amorim reveals that:

> Using different methods is a strategy for making the most of fieldwork. Studying the object from different points of view, making the analysis strategy more flexible, seems to be more efficient than choosing a single, standard method (2006, p. 15).

Emphasise,

> Although there are different understandings of fieldwork, its importance in the construction of knowledge not only in geography, but also in other sciences, is notorious. It is currently being valued and increasingly applied not only in scientific circles, but also in schools (AMORIM, 2006, p. 15).

Matheus argues that "it is clear that Geography has always made use of field activities, but it is important to emphasise that the different perspectives and appropriations demarcated throughout its practice will reflect in the consolidation of different methodological trajectories" (2005, p. 32).

Schaffer (1998 apud Amorim, 2006, p. 17), considers that:

> In reading the landscape, fieldwork is an important practice for learning geography. It effectively allows knowledge to be built from the reality that is observed, analysed and contextualised (in time and space). It is also a way of overcoming the fragmentation of knowledge, insofar as the study of reality presents a multiplicity of aspects that point to the concurrence of different areas of knowledge. Above all, it is an experience capable of providing an opportunity for the concrete and simultaneous confrontation of theory and practice.
>
> Fieldwork is a very rich and important teaching resource for the learning process, because it allows theoretical knowledge to be related and broadened to reality, providing more contextualised and dynamic approaches to school content in the process of knowledge and education.

We understand that in order to carry out fieldwork in a given location, the teacher must bear in mind that "the student must be an active subject in their learning, acting in the construction and understanding of their knowledge, and not a passive one who is limited to memorising theoretical knowledge or acquiring skills and techniques" (COMPIANI, 1991, p. 3).

According to Compiani, "we have no doubt that the field is an excellent teaching environment and, if worked well, is capable of questioning the traditional classroom, enclosed by four walls with a teacher in an inaccessible, distant position" (1991, p. 18).

To corroborate this, "the work with the field that we are proposing shows, at the very least, the need to criticise the classroom space and to break with the monotony and exclusive possession of the discourse by the teacher" (COMPIANI, 1991, p. 18).

We understand that fieldwork is a way of organising teaching that provides a coherent and consistent view of objects, phenomena and geographical processes in their environment. This underpins and gives light to direct observation, stitching together the abstract world with everyday reality.

Professor Mapatse adds: "But this is not naïve or passive observation. It is an observation that is problematised by the geographical issues of a natural, social, political and economic nature that surround these realities" (2006, p. 22).

From this perspective, I emphasise that "the teacher, by helping the student to learn and construct knowledge, is breaking away from the content-based approach" (KRAEMER, 2003, p. 9).

According to Pinto (2003, p. 17), "fieldwork is a technique, and if it is not to be just another common technique, it must surprise the learner and provoke curiosity". He goes on to say:

> Fieldwork can be one of many (special) ways of promoting the learning process. For this reason, when it enables students to come into contact with an environmental phenomenon, event or process, it must not be limited to recognising them on a one-off basis, nor must it turn one-off observations into the only ways of understanding the dynamics of a given landscape. It should prioritise interest, effective participation, dialogue and reflection (PINTO, 2003, p. 15).

It also appears that:

> Excursions help to raise the expectations of Geography in its development as a Science, because as it is social, it does not lend itself to laboratory experiments to test phenomena, and therefore excursions will serve to realise these experiments by allowing the "in loco" observation of phenomena in their natural environment (MAPATSE, 2006, p. 23).

It is also emphasised that "the geographical excursion is a form of active teaching, as it allows learning that starts from the student's needs and not just from the demands of the syllabus" (MAPATSE, 2006, p. 27).

It should be said that "it is a complement to geography teaching methods since it stimulates and increases the students' activities in the classroom, boosts their independent work and reduces their dependence on the teacher" (MAPATSE, 2006, p. 31).

For Graves (1978 apud Mapatse, 2006, p. 29),

> Geographical excursions consist of simple exercises working in the student's everyday environment, where, for example, using a road map they can observe and describe in their diary the geographical aspects that characterise the landscape along their route from home to school and vice versa.

Mapatse states that:

> Geographical excursions open up the possibility of linking theory and practice, between the real and the abstract, developing the habit of observation and permanent vigilance towards our surroundings, expanding not only cognitive capacities, but also those of applying knowledge to the demands of everyday life in the community (2006, p. 35).

According to Oliveira, "students need Geography to give them the tools to understand the world of their time and act in it" (2005, p. 9).

He goes on to point out, "the pedagogical resource routinely used in the teaching of Geography, aimed at the production of spatial knowledge and its learning, fieldwork has been conceived as an important means of achieving this objective". It has also "been used as a way of working collectively with various areas of knowledge" (OLIVEIRA, 2005, p. 9).

Thus, "it can be said that today fieldwork is more present than ever in research of interest to Geography. It is valued according to the research objectives, sometimes with classical, sometimes neopositivist, critical or humanist characteristics" (OLIVEIRA, 2005, p. 52).

Carlos (2002 apud Oliveira, 2005, p. 53) goes so far as to say that today there is "a return to empiricism in Geography and that this return is overlapping with theoretical reflection".

We need to be cautious about the above statement by Carlos (2002 apud Oliveira, 2005, p. 53) so as not to trivialise fieldwork, because,

> It can be seen that fieldwork is of great relevance to geographical studies as a research tool both in the academic environment and as a teaching tool, considering the place it has occupied throughout the construction of geographical thought (OLIVEIRA, 2005, p. 53).

Still on the subject,

> In Geography, you can't look at the landscape like a tourist who appreciates what is beautiful and criticises what is not. The geographer's gaze must be attentive to the relationships between natural elements and human occupation, seeing a geographical explanation for each singularity of spatial occupation. The field is the geographer's laboratory and his gaze his main working tool (LEME, 1999, p. 42).

So,

> Considering the period of significant transformations in which we live, the need for new ways of thinking and living and the revalorisation of space as a place of life, I understand that fieldwork enables different readings and reflection/action on geographical space (BRAUN, 2005, p. 12).

Like this,

> From this perspective, we see fieldwork as a methodological path that enables links between the various fields of Geography and between the different areas of knowledge, with the aim of contributing to the education of the citizen of the 21st century and to understanding today's world (BRAUN, 2005, p. 69).

In this situation, Callai and Zarth (1988, p. 35) consider an excursion to be "an activity that allows direct observation of a particular place".

According to the same authors (1988, p. 36),

> Fieldwork is a very interesting resource to develop with students and is a very concrete possibility when studying the municipality. Depending on the objectives you want to achieve, it can be a very quick activity, or much longer and more complex, and can be carried out using one or more techniques, such as: a survey and bibliographical research on the subject, analysis of existing material, excursions, observation, questionnaires, interviews, collection of existing material, lectures, visits, reports, construction of plans, models, exhibitions, debates, round tables, discussions, dramatisations, etc.

We understand that studying the place gives us physical access to everything that interests us. In this way, the municipality becomes significant geographical content for fieldwork. According to Callai and Zarth (1988, p. 11),

> Studying the municipality is important and necessary for students, as they are developing the process of understanding and criticising the reality in which they live. There, space and time are delimited, allowing all aspects of the complexity of the place to be analysed.

From this perspective, studying the municipality has at least two advantages:

> The first is that students are able to recognise themselves as citizens in a reality that is their real life, appropriating information and understanding how social relations and the construction of space take place. The other advantage is pedagogical, because by studying something that is experienced by the student, there is a much greater chance of success, of it becoming more consistent learning (CALLAI, 1998, p. 79).

It should also be said.

> The municipality is the smallest hierarchical political-administrative centre within the Brazilian structure. As no place can explain itself, it is necessary to constantly exercise theorisation, establishing links and seeking explanations at regional, national and even international level (CALLAI, 2000, p. 124).

"In order to do this, in addition to theory and geotechnological paraphernalia, it is necessary to go into the field to understand landscapes and their nuances, frameworks, horizons and contextualisations" (CARNEIRO, 2007, p. 23).

So,

> I say that when I savour some of Claval's ideas (THE LANDSCAPE OF THE GEOGRAPHERS), I say that you have to know how to open the window, have a clinical and sensitive eye, broaden your horizons, your perceptions and contextualise the landscapes. That's why geologist Nicolas Desmarest said: GO AND SEE (CARNEIRO, 2007, p. 24).

According to Carneiro, "the practice of fieldwork means valuing the discovery of the place" (2007, p. 42). Still for Carneiro,

> The action of planning in each branch of Physical Geography provides an orientation of "not being tied down" to the classroom, but rather fostering the practice of looking at geography through field lessons, especially focussing on our municipality. The purpose of this activity is to enable students to make contact with and reflect on municipal studies and to identify the mechanisms of man's relationship with nature (2006, p. 128).

Continuing from the municipal perspective, "field classes provide, from a theoretical foundation, the development of the interrelationship with the elements of the physical environment, underpinning the theory/practice relationship as a strong learning link" (CARNEIRO and DIAS, 2005, p. 138).

Reinforcing this line of thought, Callai (1999, p. 12) states that: "understanding reality through Geography means being able to handle the basic concepts and the appropriate instruments for carrying out research and presenting its results".

Finally, in order to get rid of academic bureaucracy and the hardships of private life for both academics and teachers, I advocate fieldwork as a way of studying the municipality, in other words, "studying the place in order to understand the world" (CALLAI, 2000, p. 83-134).

Synthesising the contributions of the different authors, we can see that according to either the objective of the activity, its planning or the role of the teacher and students, there are different classifications for fieldwork. So we have to bear in mind that "fieldwork can't just be an opportunity to break out of the classroom routine" (RODRIGUES and OTAVIANO, 2001, p. 36), but should be seen as decisive for the development of geographical knowledge.

In this way, the classification proposed by Compiani and Carneiro (1993, p. 94-96), mentioned above in this topic, will guide our master's research to understand the conceptions of students (item 3.2.) and teachers (item 3.3.) regarding fieldwork practised in Geography undergraduate courses in south-eastern Goiás.

Continuing the outline of this research, section 3.3 will present and analyse the material collected from the students.

3.3 Presentation and analysis of the results of the students' conceptions

This section will present the results, as well as analysing the data obtained via the questionnaire, which indicate the teachers' and students' conceptions of the subject of FIELD WORK. We would like to emphasise that all the questions put to the participating teachers and students will be demonstrated by means of representative graphs as well as some verbatim transcriptions.

From the outset, it is important to emphasise what is meant by CONCEPTION. According to

Bueno (1991, p. 283), the term CONCEPTION means: "the act of being conceived or generated; generation; the faculty of perceiving; knowledge; the act of making ideas". Thus, we intend to find out the opinions of teachers and students in this research.

Graphs 6 and 7 show that among the students, the female gender is particularly prominent in Pires do Rio with 60 per cent and in Catalão with 59 per cent. This means that the female students have a preference for teaching. This is confirmed by Ludke (2000 apud Baradel, 2007, p. 46),

> For a long time, teaching, especially primary teaching, has appealed to the female contingent. Quite compatible with the nature of women's roles, as they are valued in our Western society, the teaching profession has responded in full to the need to introduce women into the labour force. On the other hand, this easy assimilation has had serious consequences for the "status" of the occupation.

Graph 5: Gender of students at UEG - Pires do Rio
Source: Research carried out from November 2007 to May 2008
Org.: CARNEIRO, V. A. - July/2008

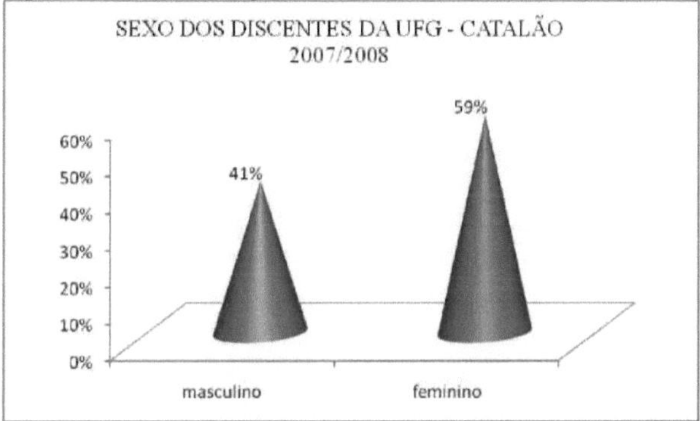

Graph 6: Sex of students at UFG - Catalão
Source: Research carried out from November 2007 to May 2008
Org.: CARNEIRO, V. A. - July/2008

Graphs 7 and 8 show a predominance of young people aged between 21 and 29, 58 per cent in Pires do Rio and 72 per cent in Catalão. We therefore conclude that these young people are seeking a place in the labour market in a more qualified way.

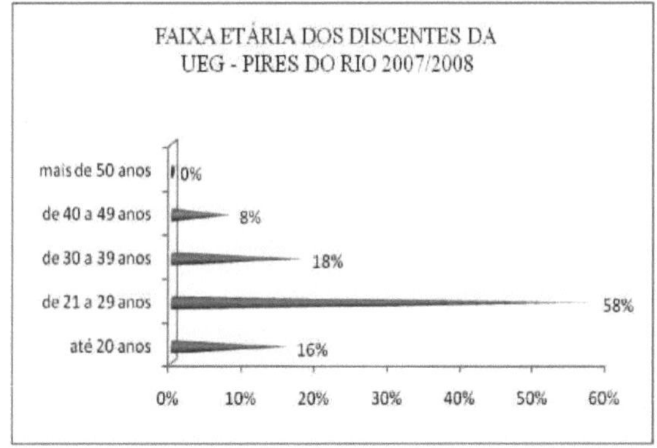

Graph 7: Age group of students at UEG - Pires do Rio Source: Survey carried out from November 2007 to May 2008 Org.

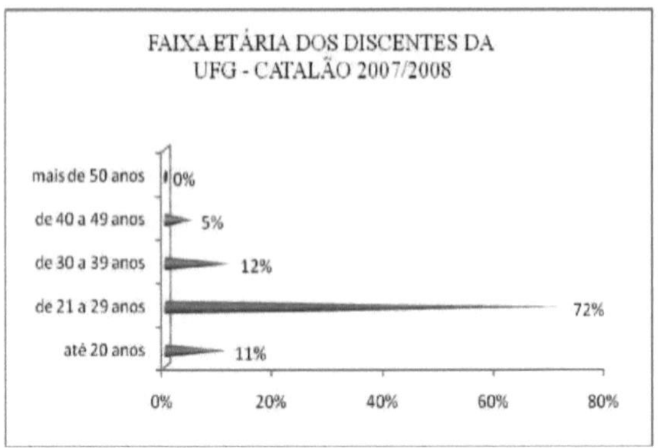

Graph 8: UFG - Catalão age group
Source: Research carried out from November 2007 to May 2008
Org.: CARNEIRO, V. A. - July/2008

Graphs 9 and 10 show that students from Pires do Rio (89%) and Catalão (98%) have strong links with the micro-region. These point to local economic activities, as mentioned above in sections 2.2.1 (Getting to know Catalão a little) and 2.2.2 (Getting to know Pires do Rio a little) of this research.

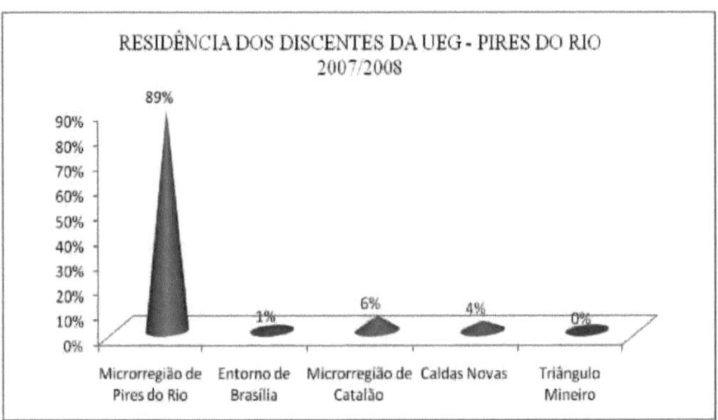

Graph 9: Residence of UEG - Pires do Rio students Source: Survey carried out from November 2007 to May 2008
Org.: CARNEIRO, V. A. - July/2008

Graph 10: Residence of UFG - Catalão students
Source: Research carried out from November 2007 to May 2008
Org.: CARNEIRO, V. A. - July/2008

Graphs 11 and 12 show that in Pires do Rio (75%) and Catalão (68%) students are working. This justifies the demand for evening courses, as in the case of the Degree in Geography, and on the other hand, there are problems with classes and fieldwork at weekends; remembering that in these municipalities in the micro-regions of Pires do Rio and Catalão there are normal commercial, industrial and agricultural activities at weekends, especially on Saturdays, which leads to a clash between academic and professional activities.

Graph 11: Working students at UEG - Pires do Rio Source: Survey carried out from November 2007 to May 2008 Org.

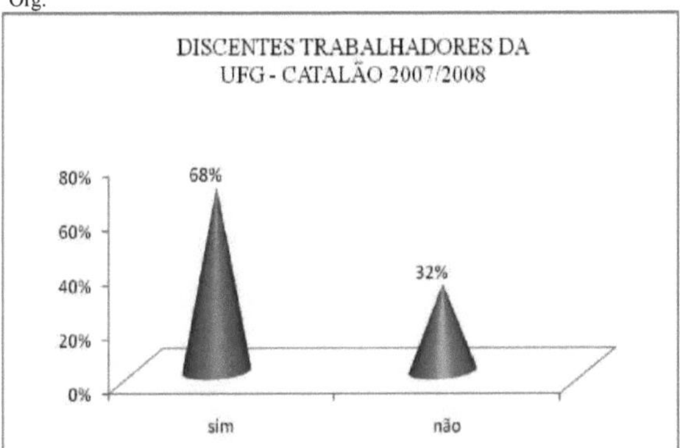

Graph 12: Working students at UFG - Catalão Source: Survey carried out from November 2007 to May 2008 Org.

Graphs 13, 14, 15 and 16 show that the majority of Geography undergraduates in the cities of Pires do Rio and Catalão come from public schools (primary and secondary). In this way, we emphasise that although public education is going through vexatious situations, there is still a light at the end of the tunnel, as long as there are coherent and consistent investments to remedy the issue of public education in our state and in our country.

Graph 13: UEG - Pires do Rio students who attended public primary schools
Source: Research carried out from November 2007 to May 2008
Org.: CARNEIRO, V. A. - July/2008

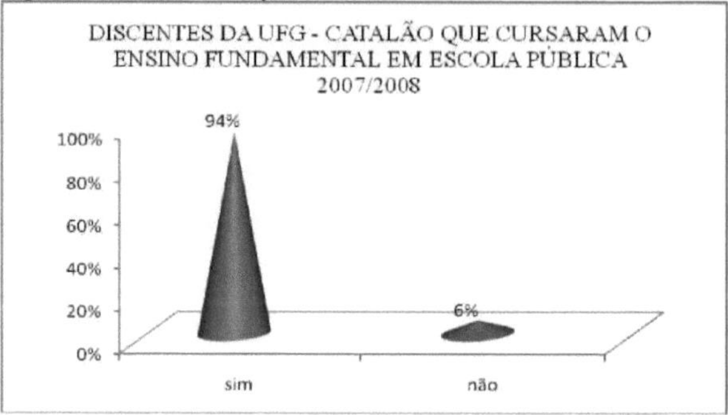

Graph 14: UFG - Catalão students who attended public primary schools
Source: Research carried out from November 2007 to May 2008
Org.: CARNEIRO, V. A. - July/2008

Graph 15: UEG - Pires do Rio students who attended public secondary schools
Source: Research carried out from November 2007 to May 2008
Org.: CARNEIRO, V. A. - July/2008

Graph 16: UFG - Catalão students who attended public secondary schools
Source: Research carried out from November 2007 to May 2008
Org.: CARNEIRO, V. A. - July/2008

In relation to graphs 17 and 18, students from Pires do Rio (36%) and Catalão (44%) show that they identify easily with the geography content taught in public schools. However, it should be emphasised that this geographical content ends up generating conflicts between school geography and university geography. Thus, we can say that university geography "condemns" traditionalism (memorisation, questionnaires, copying texts, etc.) and inserts questioning and reflection on the content practised in public schools so that its future student-teachers don't practise the same procedure as school geography:

> 'Because Geography was the course I identified with the most; I've always been interested in the subjects covered by Geography" (student 64 - Catalão).
> "Because I'm more comfortable with so-called humanities courses, such as Geography, History, etc" (student 80 - Catalão).
> "For identifying with the subject since primary school" (student 34 - Pires do Rio).
> "Because it's one of the subjects I've always found easy and identified with" (student 63 - Pires do Rio).

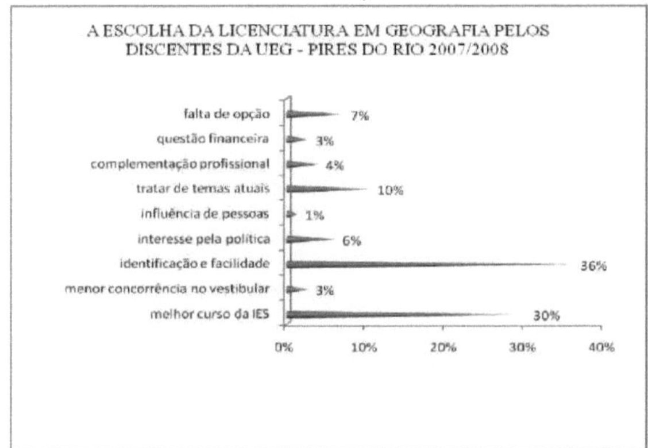

Graph 17: UEG - Pires do Rio students' choice of degree programme in Geography
Source: Research carried out from November 2007 to May 2008
Org.: CARNEIRO, V. A. - July/2008

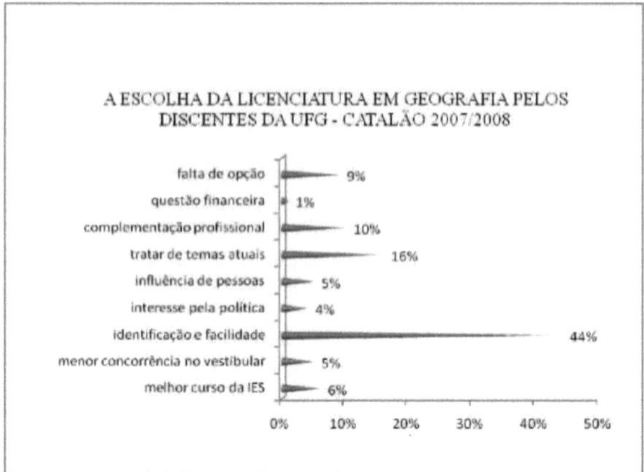

Graph 18: How UFG - Catalão students chose a degree in Geography Source: Survey conducted from November 2007 to May 2008
Org.: CARNEIRO, V. A. - July/2008

Graphs 19 and 20 show that in Pires do Rio (42 per cent) and Catalão (32 per cent), students understand that geography is linked to socio-environmental interaction. We can therefore say that Geography is based on knowing, studying and understanding the relationship between society and nature.

To this end, the students speak out:

> "It's the union and understanding of the physical and human environment" (student 48 - Pires do Rio).

> "It's an art, an art of understanding man's relationship with the environment, from the local to the global" (student 19 - Pires do Rio).
> "It's the science that studies the interactions between society and nature" (student 71 - Catalão).
> "Geography for me is the science that links the natural sciences to the science of man" (student 29 - Catalan).

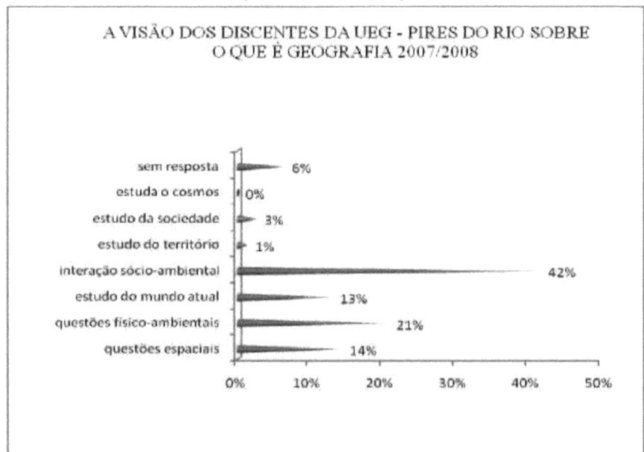

Graph 19: UEG - Pires do Rio students' view of what Geography is Source: Survey conducted from November 2007 to May 2008
Org.: CARNEIRO, V. A. - July/2008

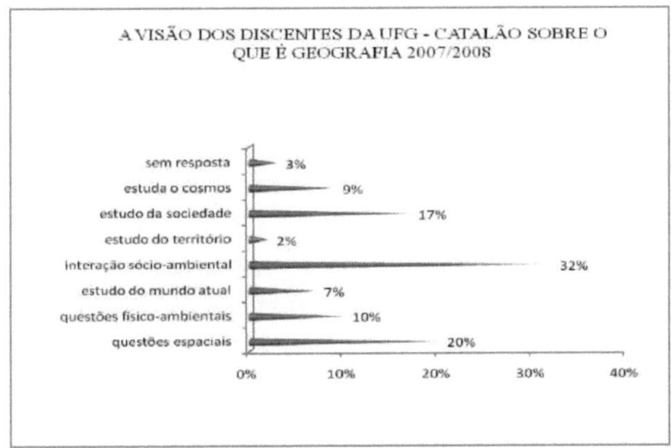

Graph 20: UFG - Catalão students' view of what Geography is Source: Survey carried out from November 2007 to May 2008
Org.: CARNEIRO, V. A. - July/2008

In Graph 21, students in Pires do Rio (30%) point out that the purpose of geography is to train critical citizens to understand and act on the current reality. In Catalão, 25% of students believe that the purpose of Geography is to understand the relationship between society and nature (Graph 22).

According to the students of Pires do Rio:

> "It serves to debate reality and get citizens to reflect and act on it" (student 59 - Pires do Rio).
> "To train critical citizens with the ability to discuss various issues" (student 58 - Pires do Rio).

For Catalan students:

> "Analysing the relationship between man and nature and proposing a balanced coexistence" (student 70 - Catalão).
> "Analysing, studying and understanding the relationship between the environment and society" (student 33 - Catalão).

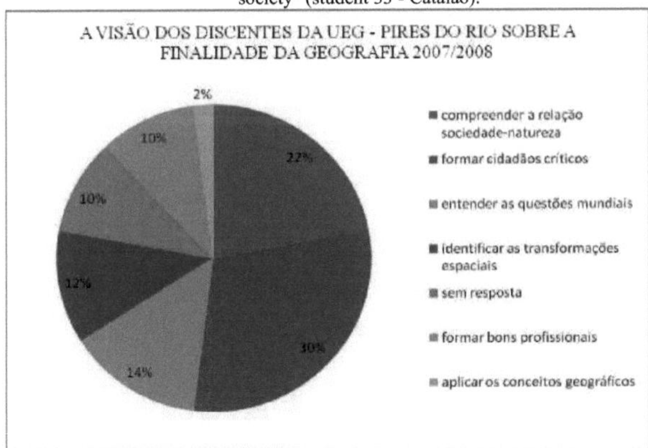

Graph 21: UEG - Pires do Rio students' view of the purpose of Geography
Source: Research carried out from November 2007 to May 2008
Org.: CARNEIRO, V. A. - July/2008

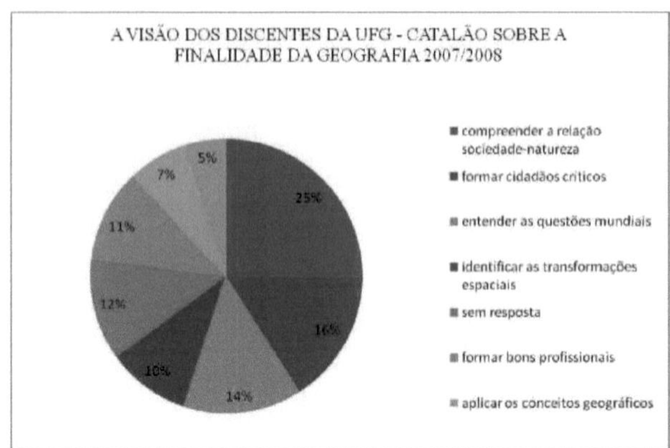

Graph 22: UFG - Catalão students' view of the purpose of Geography Source: Survey carried out from November 2007 to May 2008
Org.: CARNEIRO, V. A. - July/2008

For this question, graphs 23 and 24 show the opinion of students from Pires do Rio and Catalão, 95% and 93% respectively, agreeing that the geographical content taught at the two universities (UEG and UFG) is related to their day-to-day lives. We can therefore infer that these favourable percentages indicate a good relationship between teaching and learning.

The students justify it:

> "Yes, it shows us all the social dynamics, the geopolitics of countries, the relationship between economies, disrespect for the environment, social inequalities and the class struggle" (student 78 - Catalan).
> "Yes, because they talk about the environment I live in" (student 2 - Catalão).
> 'When we need to go to some point or city, we need to know the location, and that's Geography" (student 46 - Pires do Rio).
> "Before studying Geography, I saw nature in one way, without showing much interest; today I see it with different eyes, I observe the components of nature more" (student 66 - Pires do Rio).

Graph 23: UEG - Pires do Rio students' opinion on the relationship between geographical content and their daily lives
Source: Research carried out from November 2007 to May 2008
Org.: CARNEIRO, V. A. - July/2008

Graph 24: UFG - Catalão students' opinion on the relationship between geographical content and their daily lives
Source: Research carried out from November 2007 to May 2008
Org.: CARNEIRO, V. A. - July/2008

The students at both institutions, even though most of them are workers (graphs 11 and 12 seen above), still find a way, a "way" to take part in the fieldwork carried out by the lecturers at weekends (graphs 25 and 26).

To this question, the students say:

> "I love fieldwork because it helps us learn more quickly and accurately" (student 9 - Pires do Rio).
>
> "I participate by going to the field work, listening and always asking questions" (student 4 - Pires do Rio).
> "Because they are important for intellectual development" (student 53 - Catalão).
> "Geography has a broad scope and is present in everyday life; and the geographer is attentive to interpreting space, relationships and diagnosing and, above all, must propose solutions" (student 79 - Catalão).

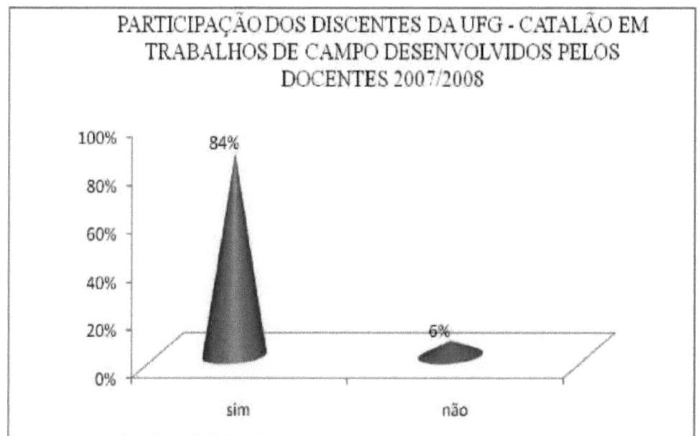

Graph 25: Participation of UFG - Catalão students in field work developed by teachers
Source: Research carried out from November 2007 to May 2008
Org.: CARNEIRO, V. A. - July/2008

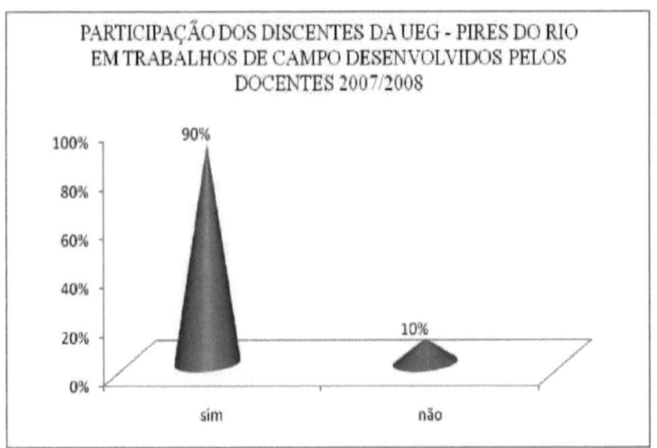

Graph26: Participation of UEG - Pires do Rio students in field work developed by teachers
Source: Research carried out from November 2007 to May 2008
Org.: CARNEIRO, V. A. - July/2008

Analysing graphs 27 and 28, we can see that students in both Pires do Rio (59%) and Catalão (67%) agree that Physical Geography subjects make much more use of fieldwork to support their content.

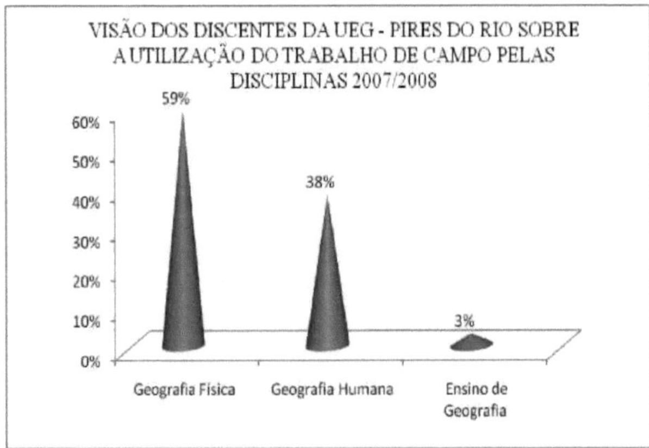

Graph 27: UEG - Pires do Rio students' views on the use of fieldwork in subjects
Source: Research carried out from November 2007 to May 2008
Org.: CARNEIRO, V. A. - July/2008

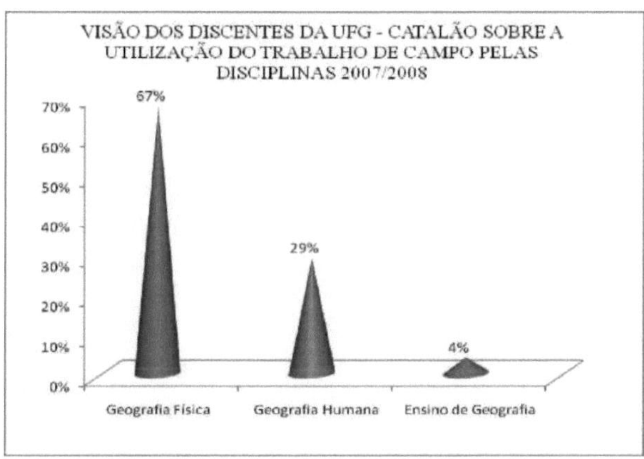

Graph 28: UFG - Catalão students' views on the use of fieldwork in subjects
Source: Research carried out from November 2007 to May 2008
Org.: CARNEIRO, V. A. - July/2008

Graphs 29 and 30 show that student participation in work in the state of Goiás are more frequent. In Pires do Rio, the frequency is 95 per cent and in Catalão it is 70 per cent. This can be explained by the fact that students negotiate their field trips with their employers.

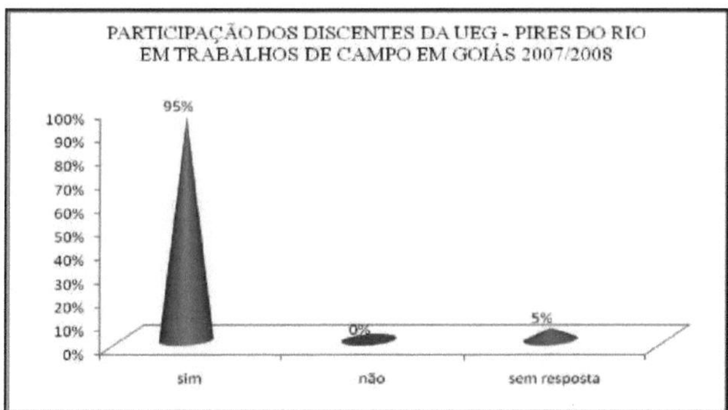

Graph 29: Participation of UEG - Pires do Rio students in fieldwork in Goiás
Source: Research carried out from November 2007 to May 2008
Org.: CARNEIRO, V. A. - July/2008

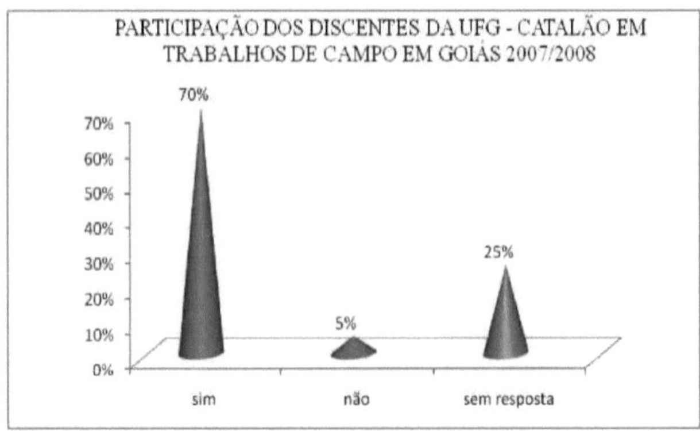

Graph 30: Participation of UFG - Catalão students in fieldwork in Goiás
Source: Research carried out from November 2007 to May 2008
Org.: CARNEIRO, V. A. July/2008

Graph 31 shows that 90% of students from Pires do Rio do not take part in fieldwork outside the state of Goiás, as there are family and work-related obstacles. Graph 32 shows that 43% of Catalão students take part in fieldwork outside Goiás.

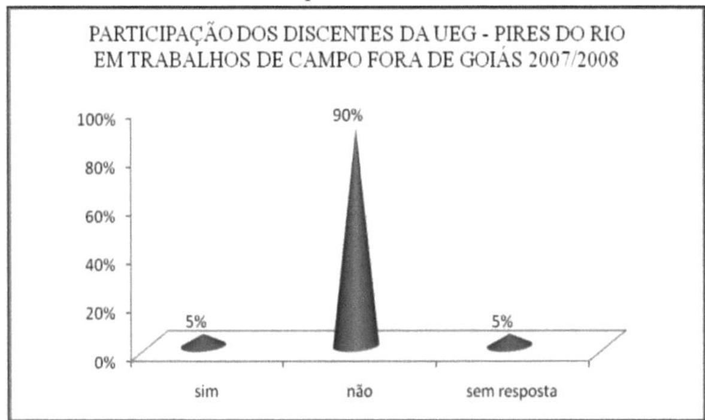

Graph 31: Participation of UEG - Pires do Rio students in fieldwork outside Goiás
Source: Research carried out from November 2007 to May 2008
Org.: CARNEIRO, V. A. - July/2008

Graph 32: Participation of UFG - Catalão students in fieldwork outside Goiás
Source: Research carried out from November 2007 to May 2008

Org.: CARNEIRO, V. A. - July/2008

Graphs 33 and 34 show that students in Catalão (59 per cent) and Pires do Rio (77 per cent) confirm that fieldwork is only carried out to assimilate content taught in the classroom. This perspective of fieldwork is traditionalist and only aims to illustrate the content taught by the lecturer.

The students are categorical when they say:

> "Fieldwork is of fundamental importance in the Geography course, as it makes it easier to understand what has been studied in class, and we can visualise in practice what we have learned in books and from our teachers" (student 60 - Catalão).
> "Learning in practice what you see in theory" (student 77 - Catalão).
>
> "Fieldwork is a chance to practise what you've studied in class" (student 76 - Catalão).
> "To see what we study in class up close" (student 3 - Catalão).
> "The importance of the field lesson is that in the field we can prove what we have learnt in the classroom" (student 69 - Pires do Rio).
> "Because we learn in practice what we study in the classroom" (student 70 - Pires do Rio).
> "For practical knowledge of the content studied in the classroom" (student 7 - Pires do Rio).
> "I think fieldwork is essential, because for me Geography isn't just about the classroom" (student 2 - Pires do Rio).

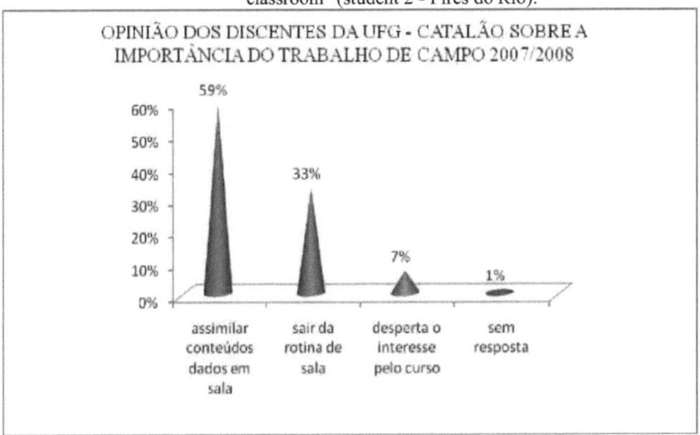

Graph 33: UEG - Pires do Rio students' opinion on the importance of fieldwork
Source: Research carried out from November 2007 to May 2008
Org.: CARNEIRO, V. A. - July/2008

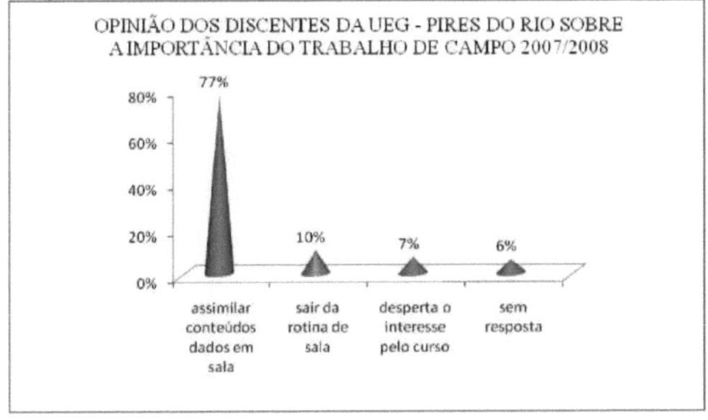

Graph 34: UFG - Catalão students' opinion on the importance of fieldwork
Source: Research carried out from November 2007 to May 2008
Org.: CARNEIRO, V. A. - July/2008

Analysing graphs 35 and 36, the difficulties faced by students in Pires do Rio (37%) and Catalão (39%) in carrying out fieldwork are the same, i.e. there is a lack of infrastructure and support for such educational activities. So we can say that this is not just a situation in Pires do Rio and Catalão, but in many public higher education institutions that are in a state of disrepair.

Regarding the difficulties in carrying out the fieldwork, the students reported:

> "Lack of help from the University Unit" (student 11 - Pires do Rio).
> "Transport and accommodation" (student 62 - Pires do Rio).
> "The lack of subsidies from UFG" (student 56 - Catalão).
> "Lack of financial support from the University" (student 54 - Catalão).

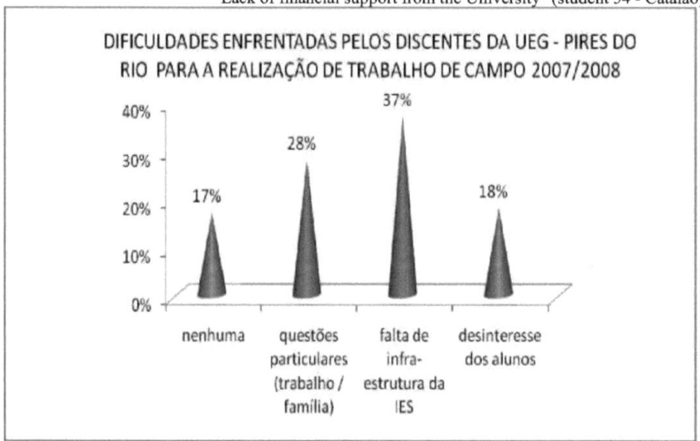

Graph 35: Difficulties faced by students at UEG - Pires do Rio when carrying out fieldwork
Source: Research carried out from November 2007 to May 2008
Org.: CARNEIRO, V. A. - July/2008

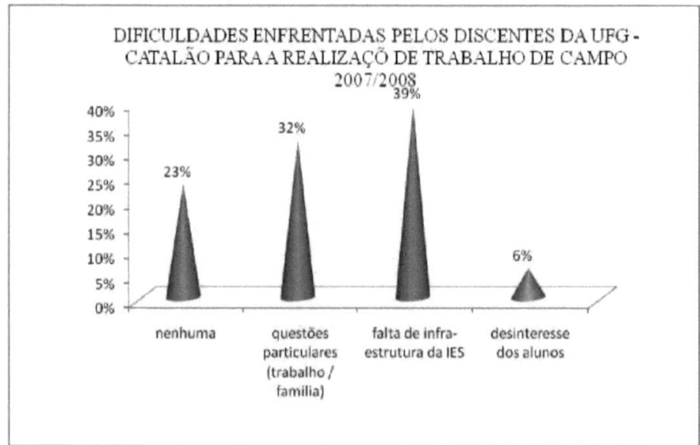

Graph 36: Difficulties faced by UFG - Catalão students when carrying out fieldwork
Source: Research carried out from November 2007 to May 2008
Org.: CARNEIRO, V. A. July/2008

Graphs 37 and 38 show that students from Pires do Rio (44 per cent) and Catalão (60 per cent) prefer to carry out fieldwork at weekends, as these are very close to each other and within the area covered by the Pires do Rio and Catalão micro-regions.

Graph 37: Periods set aside for fieldwork by UEG - Pires do Rio students
Source: Research carried out from November 2007 to May 2008
Org.: CARNEIRO, V. A. July/2008

Graph 38: Periods set aside for fieldwork by UFG - Catalão students
Source: Research carried out from November 2007 to May 2008
Org.: CARNEIRO, V. A. July/2008

When examining student participation in fieldwork projects (teaching, research and extension), we noticed high percentages in Pires do Rio (91%) and Catalão (85%) who do not participate or are unaware of such projects. Generally, these projects are designed for a single purpose, i.e. to develop complementary activities within a subject (graphs 39 and 40).

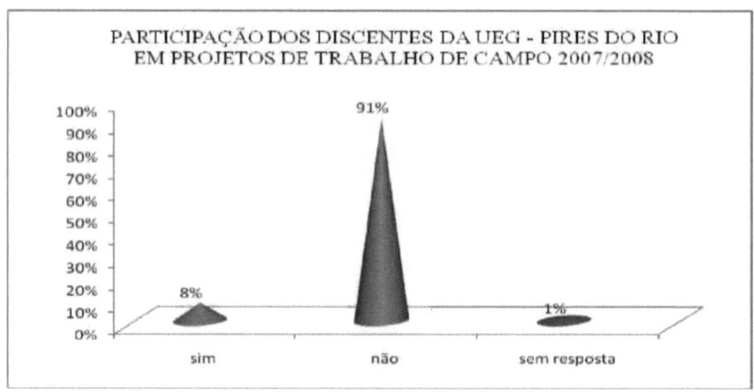

Graph 39: Participation of UEG - Pires do Rio students in fieldwork projects
Source: Research carried out from November 2007 to May 2008
Org.: CARNEIRO, V. A. - July/2008

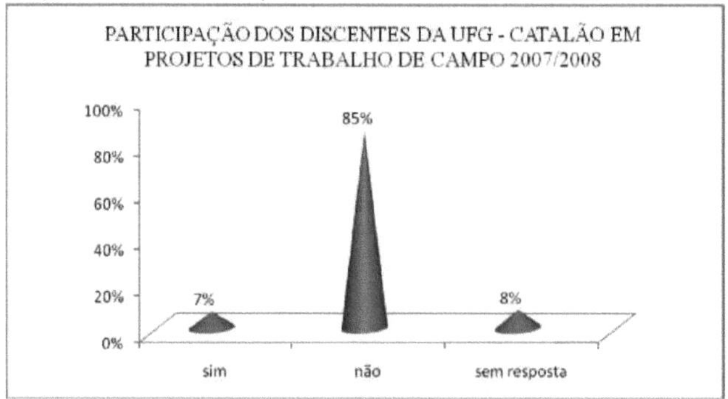

Graph 40: Participation of UFG - Catalão students in fieldwork projects
Source: Research carried out from November 2007 to May 2008
Org.: CARNEIRO, V. A. - July/2008

In Graphs 41 and 42, students from Pires do Rio (32%) and Catalão (37%) did not comment on the negative aspects of fieldwork. In this case, we turn to the data from the second post, where students from UEG and UFG, 23% and 25% respectively, say that there are no negative points about fieldwork.

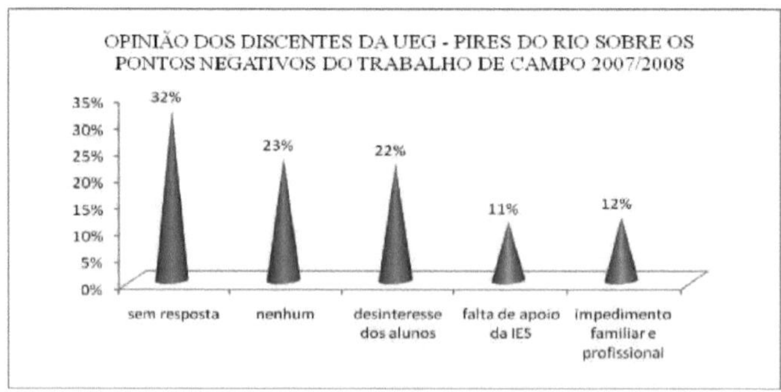

Graph 41: UEG - Pires do Rio students' opinions on the negative aspects of fieldwork
Source: Research carried out from November 2007 to May 2008

Org.: CARNEIRO, V. A. - July/2008

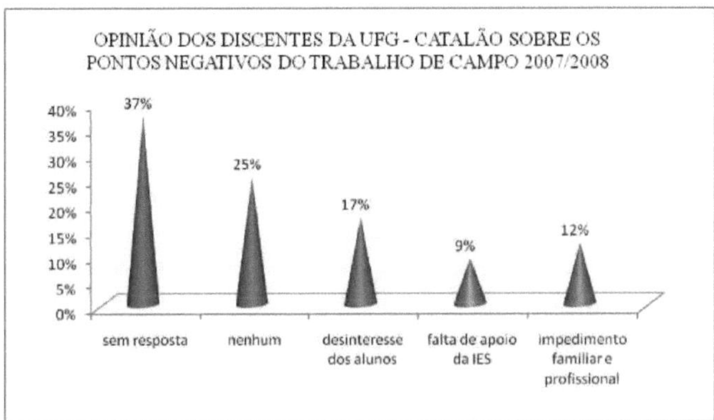

Graph 42: UFG - Catalão students' opinions on the negative aspects of fieldwork
Source: Research carried out from November 2007 to May 2008
Org.: CARNEIRO, V. A. - July/2008

With regard to the positive aspects of fieldwork, graphs 43 and 44 show that students from Pires do Rio (51 per cent) and those from Catalão (40 per cent) describe it as beneficial in terms of retaining the geographical content covered in the classroom.

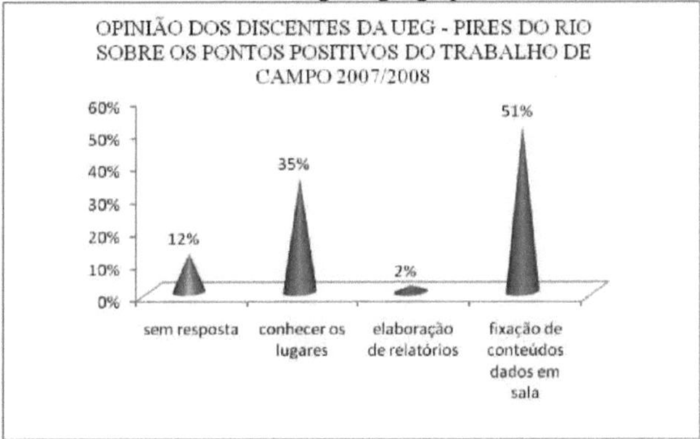

Graph 43: UEG - Pires do Rio students' opinions on the positive points of fieldwork
Source: Research carried out from November 2007 to May 2008
Org.: CARNEIRO, V. A. - July/2008

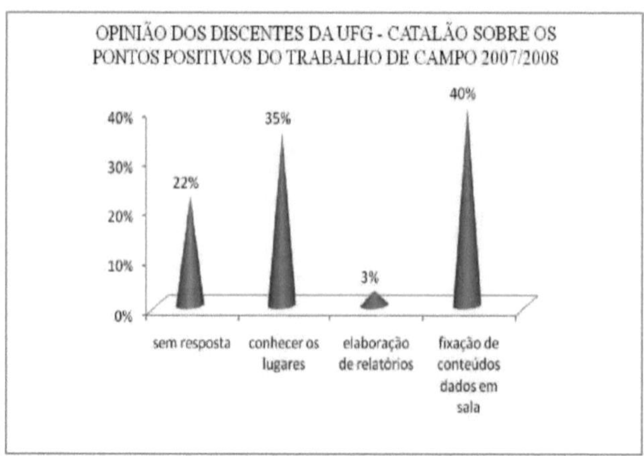

Graph 44: UFG - Catalão students' opinions on the positive points of fieldwork
Source: Research carried out from November 2007 to May 2008
Org.: CARNEIRO, V. A. - July/2008

Looking at graphs 45 and 46, we can see that there are disparities in the definitions (types) of fieldwork for students. In this case, graph 45, students from Pires do Rio (52%) indicate that fieldwork is motivational. In this respect, we can emphasise that the motivating role of fieldwork in Pires do Rio is in line with the Geography degree system. It should be noted that Pires do Rio does not have a bachelor's degree system. On the other hand, students in Catalão (29%) say that fieldwork has a training aspect. This training aspect seen by the Catalão students in relation to fieldwork is linked to the foundations of the bachelor's degree programme in Geography that many are aiming for.

The students from Pires do Rio opted for motivational fieldwork, which aims to arouse students' interest in a given problem or aspect to be studied, valuing the experience of each student and their questions. From this perspective, we can see that in Pires do Rio there is a link to teacher training for the municipal and state education networks. The students in Catalão, on the other hand, chose training fieldwork, which aims to train skills, usually using scientific apparatus, instruments or devices. In the case of Catalão,

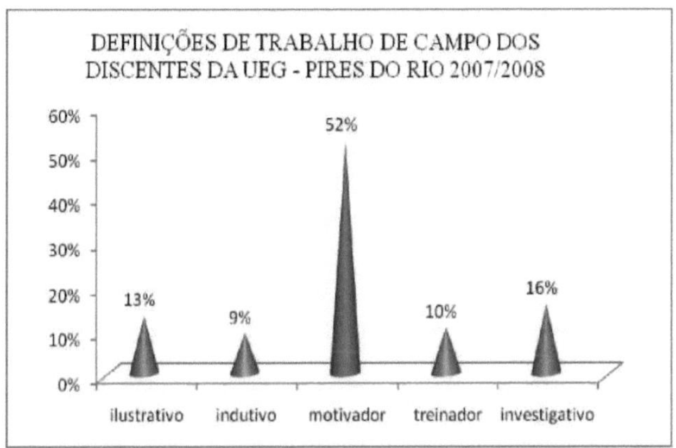

Graph 45: UEG - Pires do Rio students' definitions of fieldwork Source: Survey carried out from November 2007 to May 2008
Org.: CARNEIRO, V. A. - July/2008

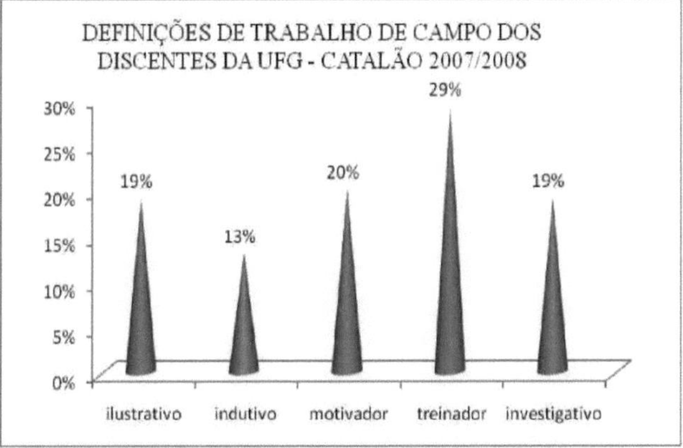

Graph 46: Questionnaires applied to students at UEG - Pires do Rio (2007-2008)
Source: Research carried out from November 2007 to May 2008
Org.: CARNEIRO, V. A. - July/2008

According to graphs 47 and 48, students in both Catalão (70%) and Pires do Rio (82%) show great interest in publicising their reports at geographical events. Therefore, we believe that this type of thinking is valid and is already starting the development of scientific production, even if it is incipient. It's worth pointing out that the way to publicise your reports at geographic events, as long as you comply with the technical standards, is in the session called experience reports.

According to Pinheiro (2005, p. 76), "the experience report modality refers to research that describes and analyses an educational practice promoted and carried out either in non-specific situations or in specific situations".

The students' opinions are:

> "It would be a way of expanding experiences, in short, a way of motivating and encouraging university students" (student 2 - Catalão).
> "Because it would show our impressions of the space" (student 28 - Catalão).
> "So that we can pass on our experience to other people" (student 7 - Catalão).
> "All fieldwork is of a scientific nature, so we should contribute to the dissemination of this knowledge" (student 60 - Pires do Rio).
> "Because we can pass on our knowledge to other people" (student 50 - Pires do Rio).
> "So that students are more motivated in their studies and projects" (student 64 - Pires do Rio).

Graph 47: Opinion of UFG - Catalão students on the dissemination of fieldwork reports at geographic events
Source: Survey carried out from November 2007 to May 2008
Org.: CARNEIRO, V. A. - July/2008

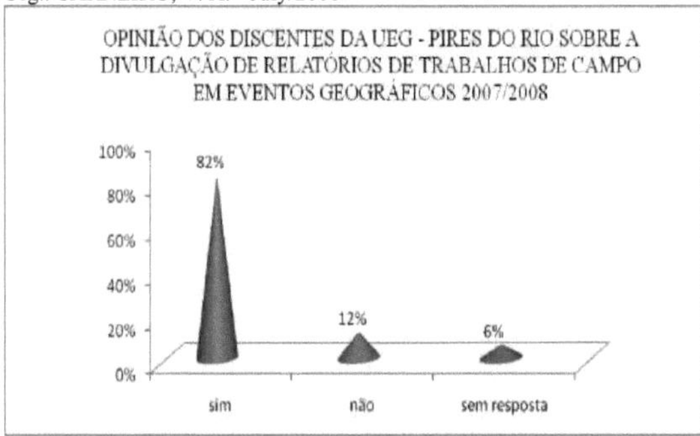

Graph 48: UEG - Pires do Rio students' opinion on the dissemination of fieldwork reports at geographic events
Source: Research carried out from November 2007 to May 2008
Org.: CARNEIRO, V. A. - July/2008

Graphs 49 and 50 show that students in Pires do Rio (62%) and Catalão (49%) did not take part in fieldwork during their secondary and primary education. In public schools (primary and secondary), there is no specific plan for developing this type of activity with students in a suitable way. Some field study activities are valid, but they are aimed at recreation, leisure and emphasise the picturesque, the bucolic. It's worth emphasising that if the environmental study technique is carried out well, it fulfils its pedagogical role.

According to Malysz (2007, p. 171), "the study of the environment enables us to perceive the action of society in space and time and also to perceive ourselves as subjects". Thus, we say that you don't have to go far, just take a look at your surroundings to realise your pedagogical actions of a geographical nature.

Here are some of the students' impressions:

> "Very simple, perhaps due to the lack of resources in public education or the lack of interest on the part of teachers" (student 8 - Catalão).
> "They were important moments for our understanding" (student 69 - Catalão).
> "It was great. However, it was on these visits that I learnt the most" (student 39 - Catalão).
> "It was very important for me, because I had an experience that helped me choose the Geography course" (student 66 - Catalão).
> "With a little fear and curiosity due to my young age" (student 55 - Catalão).
> "The work was very tied to the studies we worked on in the classroom" (student 58 - Catalão).
> "It was great, everything was new at the time" (student 77 - Pires do Rio).
> "Satisfactory, because they could be more elaborate and in-depth" (student 73 - Pires do Rio).
> "It was very good, enriching the knowledge acquired in class" (student 72 - Pires do Rio).
> "It was useful, it helped me to like Geography" (student 5 - Pires do Rio).
> "Good. It was interesting and we took the opportunity to do a cycle ride" (student 31 - Pires do Rio).
> "It was more for tourism than for the field, because the students' lack of interest made it monotonous" (student 33 - Pires do Rio).

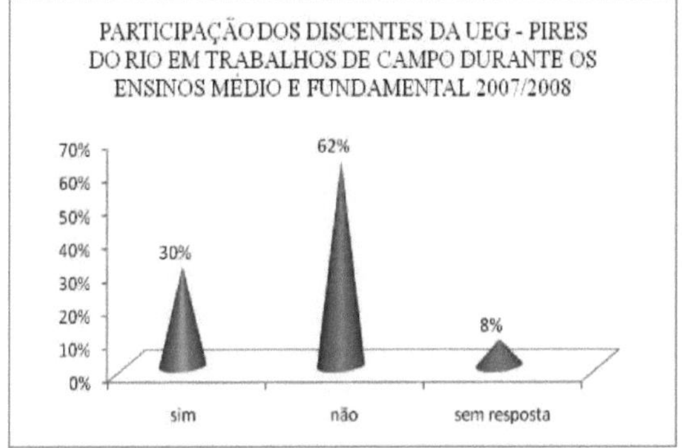

Graph 49: Participation of UEG - Pires do Rio students in fieldwork during secondary and primary education
Source: Research carried out from November 2007 to May 2008
Org.: CARNEIRO, V. A. - July/2008

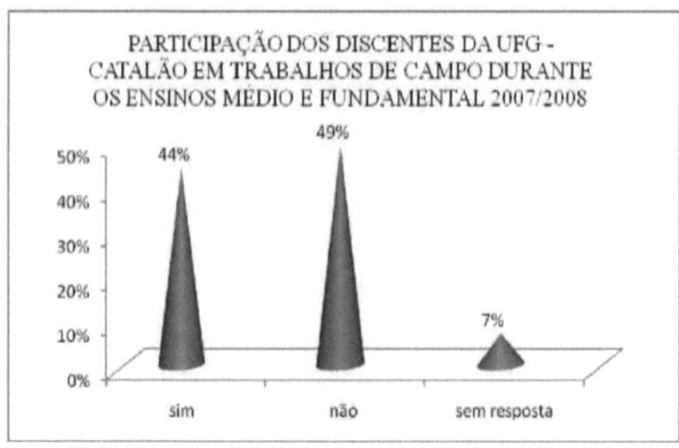

Graph 50: Participation of UFG - Catalão students in fieldwork during secondary and primary education
Source: Research carried out from November 2007 to May 2008
Org.: CARNEIRO, V. A. - July/2008

According to graphs 51 and 52, students point out that in Pires do Rio (35%) and Catalão (33%) their conceptions of fieldwork are in line with the content taught in the classroom. In this way, we can see that the teacher conducts the fieldwork from start to finish, complying with a programme that has already been stipulated according to his syllabus. It's worth noting at this point that the teacher must understand that the syllabus must not be rigid, but flexible so that the student feels like a participant in the educational process.According to the students:

> "It's very good, because it shows the student everything that was explained in class" (student 6 - Pires do Rio).
> "It contributes to learning and puts our classroom theory into practice" (student 51 - Pires do Rio).
> "Fieldwork for me is putting theory into practice" (student 53 - Pires do Rio).
> "A fundamental activity for developing practice and perception of theoretical content" (student 47 - Catalão).
> "Extremely important for understanding the theory" (student 62 - Catalão). "An activity that aims to give students the experience of witnessing the concepts put forward in theory in reality" (student 37 - Catalão).

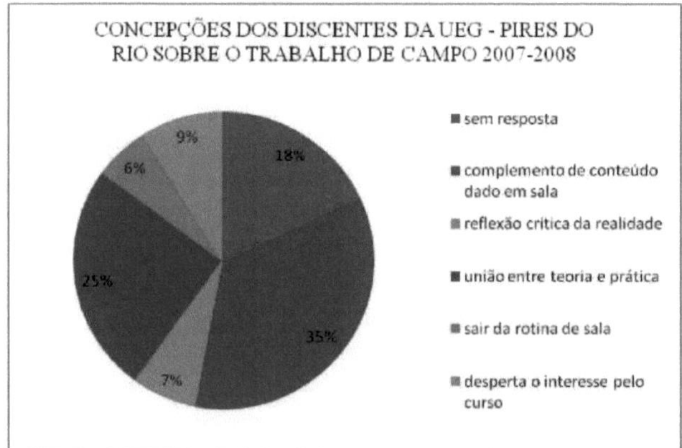

Graph 51: UEG - Pires do Rio students' conceptions of fieldwork
Source: Research carried out from November 2007 to May 2008
Org.: CARNEIRO, V. A. - July/2008

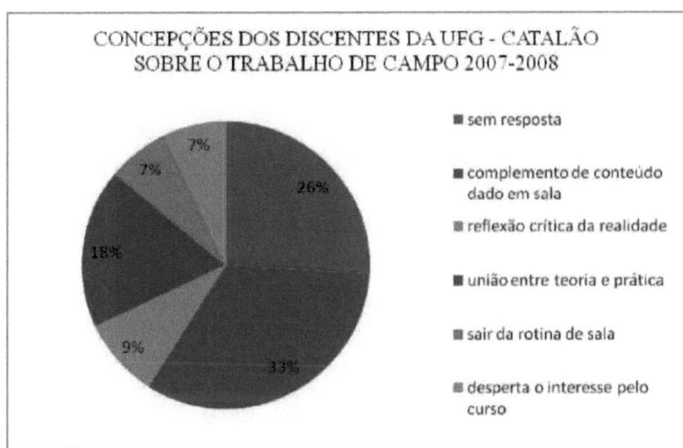

Graph 52: UFG - Catalão students' conceptions of fieldwork
Source: Research carried out from November 2007 to May 2008
Org.: CARNEIRO, V. A. - July/2008

In short, with regard to the students' conceptions, we noticed differences in graphs 21, 22, 31, 32, 45 and 46. Graph 21 emphasises the formation of critical citizens as the basis for the purpose of Geography. Graph 22 informs us that the purpose of Geography is based on understanding the society-nature relationship. Graph 31 shows that Pires do Rio students do not participate in fieldwork outside the state of Goiás. Graph 32 shows a medium level of participation by students in fieldwork outside Goiás. Graph 45 shows that motivational fieldwork is the most common in Pires do Rio, while Graph 46 shows that coaching fieldwork is the most common in Catalão.

Continuing the work, we will now look at the teachers' conceptions of fieldwork.

3.4 Presentation and analysis of the results of teachers' conceptions

At this stage of the research, in addition to the views of the students (item 3.3), it is also necessary to find out the opinions of the teachers at the two higher education units.

Graphs 53 and 54 show a predominance of tenured teachers in both Pires do Rio (73 per cent) and Catalão (89 per cent). The other percentages refer to teachers working on temporary contracts.

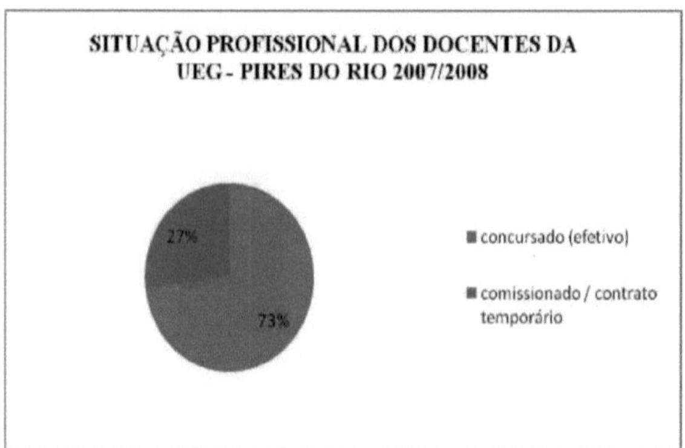

Graph 53: Professional status of teachers at UEG - Pires do Rio Source: Survey conducted from November 2007 to May 2008
Org.: CARNEIRO, V. A. - July/2008

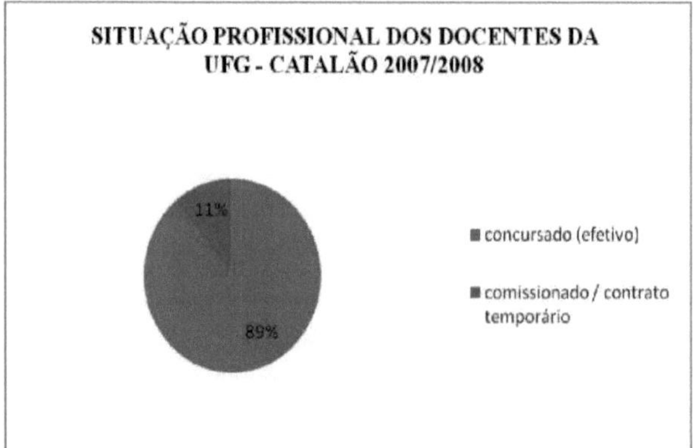

Graph 54: Status of teachers at UFG - Catalão Source: Survey carried out from November 2007 to May 2008 Org.

Looking at Graph 55, we can see that the length of time teaching at Pires do Rio is at two different levels: five years (37%) and 10 years (37%). This is due to the young age of the institution, as well as the fact that it recently had a public competition and the high turnover of teachers in search of new opportunities and better salaries at other institutions.

Graph 56 shows a predominance of teachers with 25 years' service (33 per cent), while there are also two groups ranging from 10 years (22 per cent) to 15 years (23 per cent). The justification for this is based on the academic maturity acquired by the institution in Goiás.

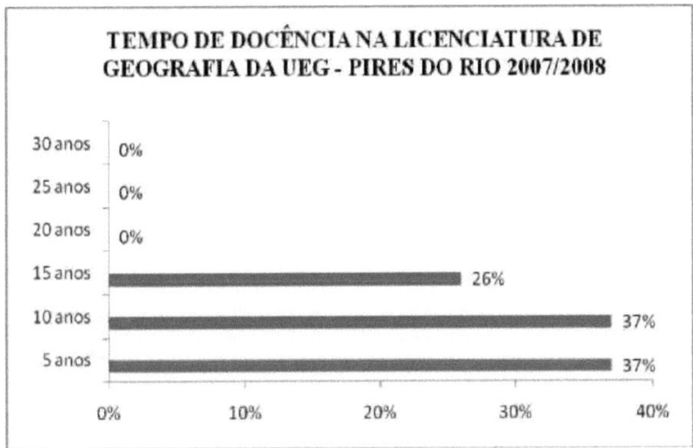

Graph 55: Length of time teaching Geography at UEG - Pires do Rio
Source: Research carried out from November 2007 to May 2008
Org.: CARNEIRO, V. A. - July/2008

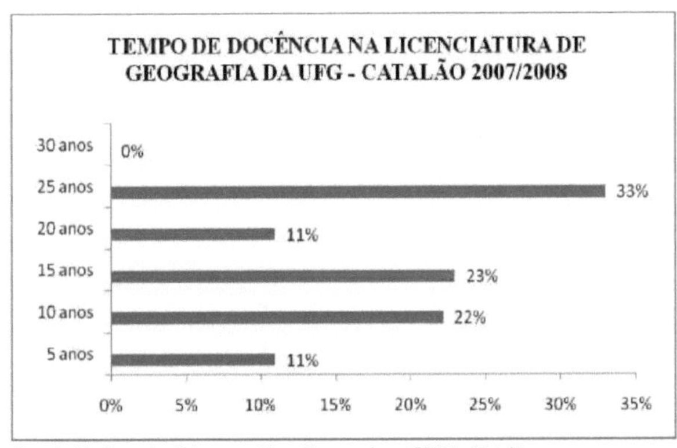

Graph 56: Length of time teaching Geography at UFG - Catalão
Source: Research carried out from November 2007 to May 2008
Org.: CARNEIRO, V. A. - July/2008

Graphs 57 and 58 show that in Pires do Rio (82%) and Catalão (100%), teachers use fieldwork in their subjects for various purposes. This shows the presence of this activity within the Geography degree programmes at both institutions.

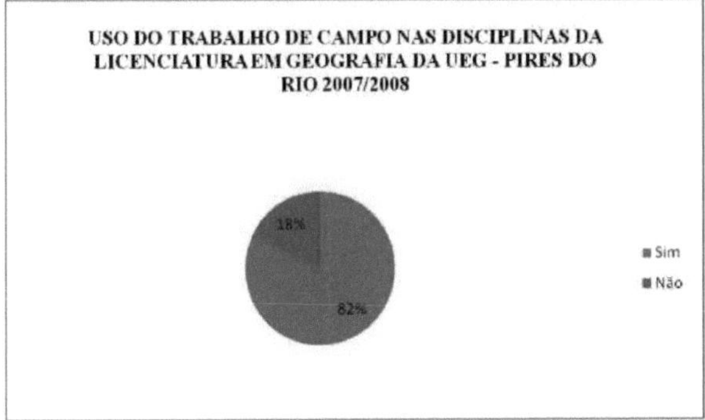

Graph 57: Use of fieldwork in Geography degree courses at UEG - Pires do Rio
Source: Research carried out from November 2007 to May 2008
Org.: CARNEIRO, V. A. - July/2008

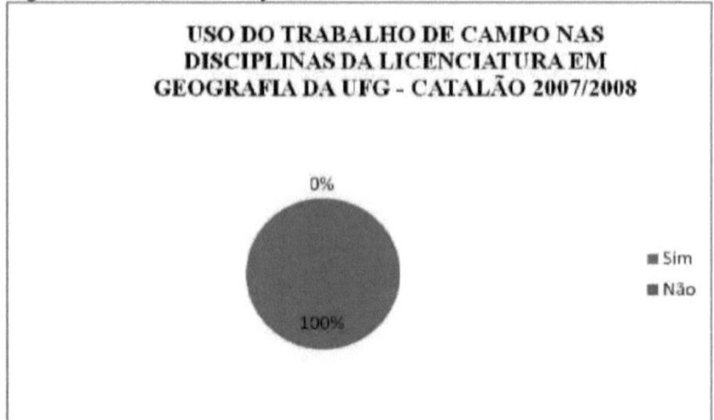

Graph 58: Use of fieldwork in Geography degree courses at UFG - Catalão
Source: Research carried out from November 2007 to May 2008
Org.: CARNEIRO, V. A. - July/2008

Teachers in Pires do Rio (82%) and Catalão (67%) prefer to carry out fieldwork in the state of Goiás, as it makes it easier for students to take part and the activities usually take place at weekends (graphs 59 and 60).

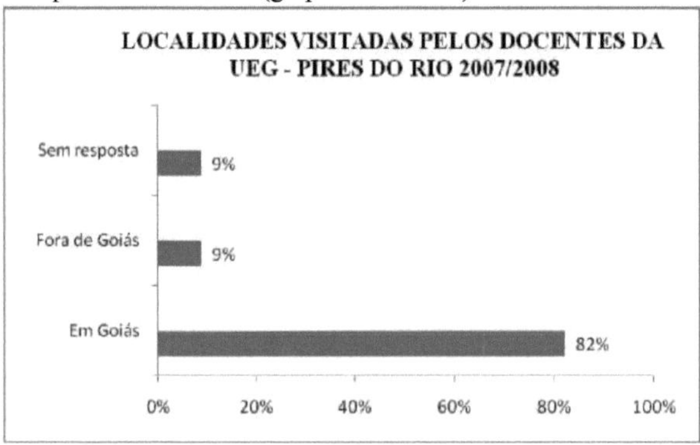

Graph 59: Localities visited by UEG - Pires do Rio teachers Source: Survey carried out from November 2007 to May 2008
Org.: CARNEIRO, V. A. - July/2008

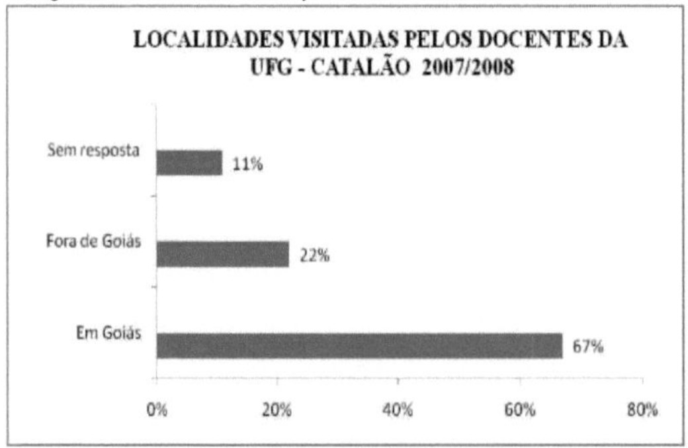

Graph 60: Locations visited by UFG - Catalão teachers

Source: Research carried out from November 2007 to May 2008
Org.: CARNEIRO, V. A. - July/2008

Regarding the importance of fieldwork, teachers in Catalão (89%) and Pires do Rio (82%) emphasise the link between theory and practice as a mechanism for pedagogical edification. Thus, we can see that fieldwork acts as a bridge between theoretical and practical knowledge in Geography degrees (graphs 61 and 62).

According to the teachers:

> "It provides opportunities to visualise and perceive aspects of the landscapes theorised in class" (Teacher 8 - Catalão).
> "It's fundamental, because it encourages dialogue between theories and social reality. I think it's very difficult to grasp geographical knowledge without field activities" (Teacher 5 - Catalão)."It's extremely important, it complements laboratory activities. Geography is done with the feet and the head" (Teacher 3 - Catalão)."Fieldwork provides a better discussion of theoretical content, it's a moment for students and teachers to experience our object of study, society integrated into the environment and the vision of all its interrelationships" (Teacher 1 - Pires do Rio)."It's very important because it brings students closer to their reality and allows them to reflect on theory and practice in the Geography course" (Teacher 3 - Pires do Rio). "Fieldwork is important because of the interaction between practice and the theory acquired by the students" (Teacher 11 - Pires do Rio).

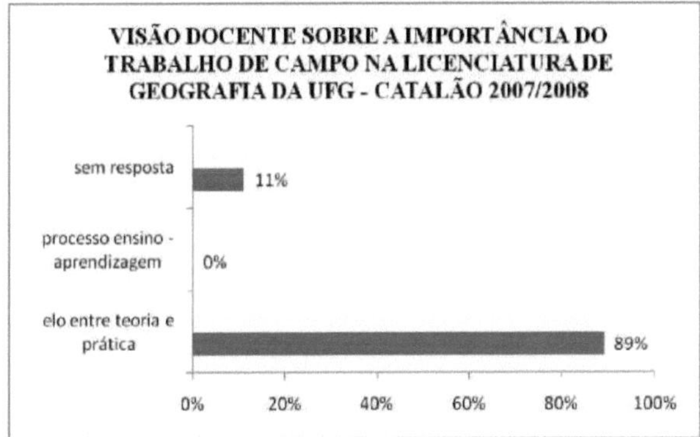

Graph 61: Teachers' views on the importance of fieldwork in the Geography degree programme at UFG - Catalão
Source: Research carried out from November 2007 to May 2008
Org.: CARNEIRO, V. A. - July/2008

Graph 62: Teachers' views on the importance of fieldwork in the Geography degree programme at UEG - Pires do

Rio
Source: Research carried out from November 2007 to May 2008
Org.: CARNEIRO, V. A. - July/2008

In graphs 63 and 64, the teachers at both Pires do Rio (83%) and Catalão (67%) show that the problems in carrying out fieldwork lie in the lack of infrastructure (buses, accommodation and other subsidies) on the part of both institutions. Even so, the mutual efforts of teachers and students try to get around this infrastructural situation so that the scheduled activity can be carried out.

The problems in carrying out the fieldwork are:

> "Funding for transport and other expenses" (Teacher 8 - Pires do Rio).
> "Transport" (Teacher 10 - Pires do Rio).
> "Expensive and precarious transport" (Teacher 1 - Catalão).
> "University resources" (Lecturer 6 - Catalão).

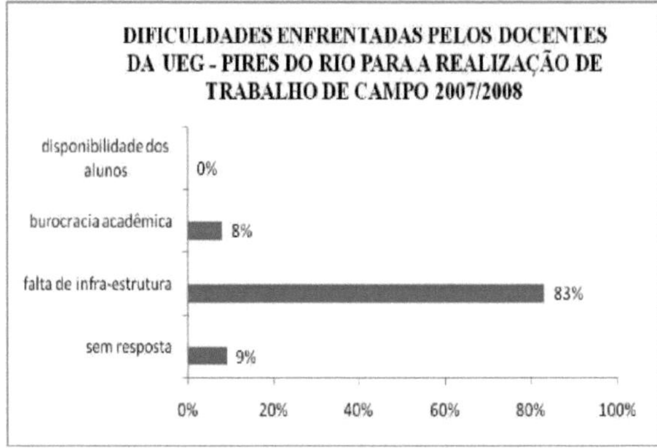

Graph 63: Difficulties faced by UEG - Pires do Rio teachers in carrying out fieldwork
Source: Research carried out from November 2007 to May 2008
Org.: CARNEIRO, V. A. - July/2008

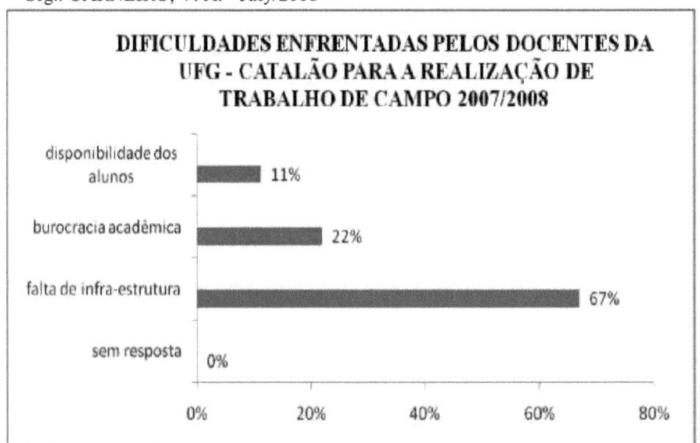

Graph 64: Difficulties faced by teachers at UFG - Catalão when carrying out fieldwork
Source: Research carried out from November 2007 to May 2008
Org.: CARNEIRO, V. A. - July/2008

Teachers in Pires do Rio (82%) and Catalão (56%) show that they frequently use fieldwork in their subjects. We can therefore infer that most of these activities end up being carried out in the municipalities themselves or in neighbouring areas to ensure their frequency (graphs 65 and 66).

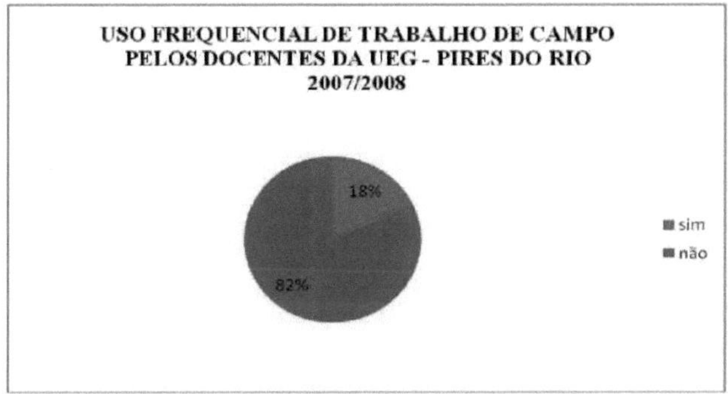

Graph 65: Frequent use of fieldwork by UEG - Pires do Rio teachers
Source: Research carried out from November 2007 to May 2008
Org.: CARNEIRO, V. A. - July/2008

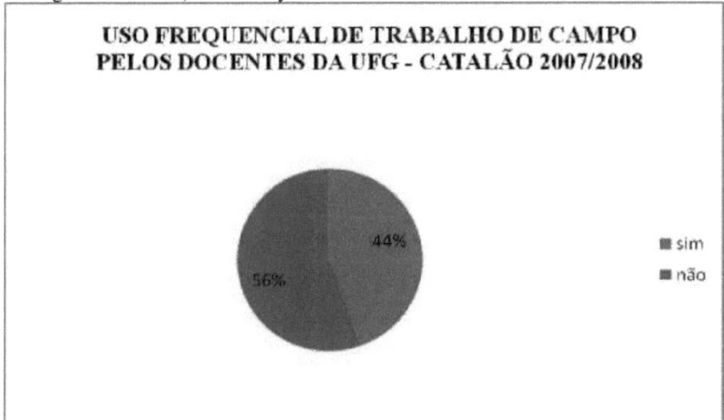

Graph 66: Frequent use of fieldwork by teachers at UFG - Catalão Source: Survey carried out from November 2007 to May 2008 Org.

According to graphs 67 and 68, the weekend stands out for fieldwork in both Pires do Rio (45%) and Catalão (68%), because it is a short period of time and aims to include more students, including those who work. In this way, students end up making themselves available so as not to miss this activity scheduled by the teachers.

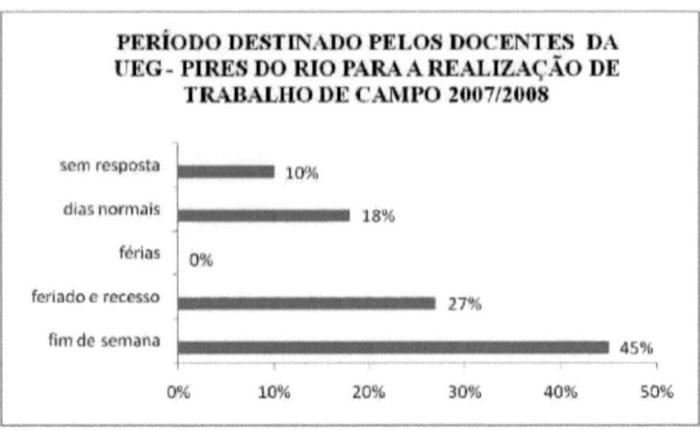

Graph 67: Period set aside by UEG - Pires do Rio teachers to carry out fieldwork
Source: Research carried out from November 2007 to May 2008
Org.: CARNEIRO, V. A. - July/2008

Graph 68: Period set aside by UFG - Catalão teachers to carry out fieldwork
Source: Research carried out from November 2007 to May 2008
Org.: CARNEIRO, V. A. - July/2008

Analysing graphs 69 and 70, we see that the teachers in Pires do Rio (82%) and Catalão (78%) do not draw up projects involving fieldwork. So we conclude that they only go into the field in connection with their respective subjects, reinforcing the pedagogical aspect of the bridge between theory and practice.

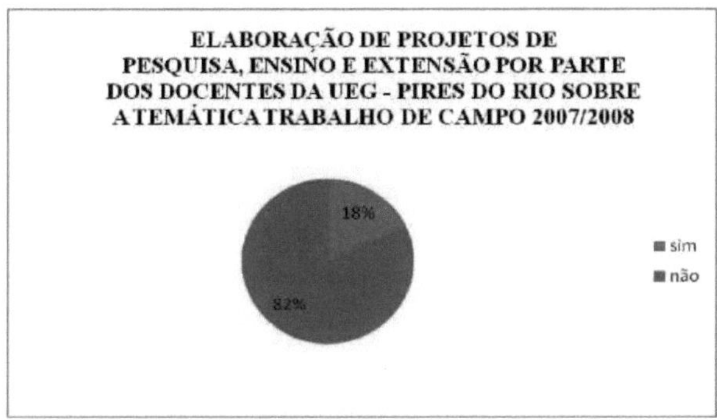

Graph 69: Elaboration of research, teaching and extension projects by UEG - Pires do Rio teachers on the subject of fieldwork
Source: Research carried out from November 2007 to May 2008
Org.: CARNEIRO, V. A. - July/2008

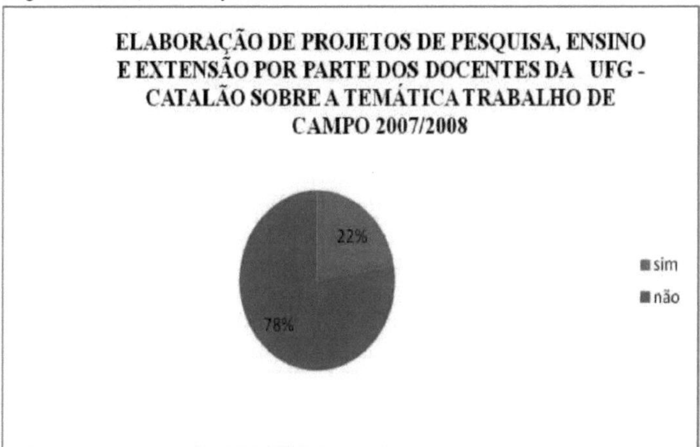

Graph 70: Development of research, teaching and extension projects by UFG - Catalão teachers on the subject of fieldwork
Source: Research carried out from November 2007 to May 2008
Org.: CARNEIRO, V. A. - July/2008

The teachers from Pires do Rio (73%) and Catalão (78%) point out that the positive points of fieldwork are based on the fact that students become more interested in, "take a liking to" the Geography degree course. From this point of view, we understand that fieldwork provides one more addition so that undecided students can define their situation in the course in question (graphs 71 and 72).On the positive side of fieldwork, the teachers made the following comments:

> "Greater learning" (Teacher 7 - Catalão).
> "Knowledge of new landscapes and places" (Teacher 2 - Catalão). "Learning for the group involved" (Teacher 4 - Pires do Rio).
> "It improves the students' assimilation of the content" (Teacher 2 - Pires do Rio).

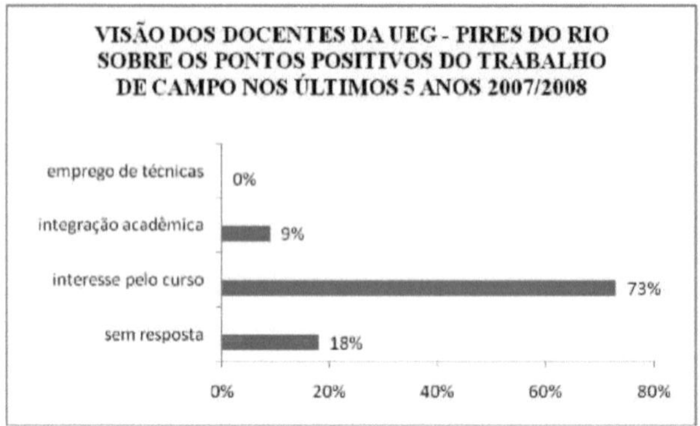

Graph 71: UEG - Pires do Rio teachers' views on the positive aspects of fieldwork in the last 5 years
Source: Research carried out from November 2007 to May 2008
Org.: CARNEIRO, V. A. - July/2008

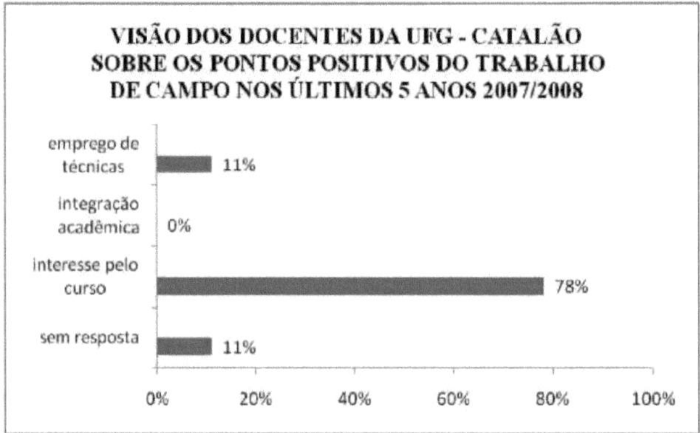

Graph 72: UFG - Catalão lecturers' views on the positive aspects of fieldwork in the last 5 years
Source: Research carried out from November 2007 to May 2008 Org.

For this approach, the teachers from Pires do Rio (36 per cent) left this item unanswered, while another group of teachers (28 per cent) reported a lack of interest on the part of students as a negative aspect of fieldwork. We realise that, in the field, if the teacher doesn't hold the reins very tightly, the goals set become a mere stroll (graph 73). Graph 74 shows a technical tie between lack of interest on the part of students (34%) and lack of transport (33%) as the negative points of fieldwork highlighted by teachers in Catalão. It's worth clarifying that the lack of transport is part of the institution's infrastructure. And the issue of student disinterest is centred on the fact that they are also displaced within the Geography course. With regard to the negative side of fieldwork, the teachers made the following comments:

"Student demotivation" (Teacher 3 - Pires do Rio).
"Little student participation" (Teacher 8 - Pires do Rio).
"Student discouragement" (Teacher 4 - Catalão).
"Lack of transport" (Teacher 2 - Catalão).
"The bus is missing" (Teacher 7 - Catalão).
"The students lack interest" (Teacher 6 - Catalão).

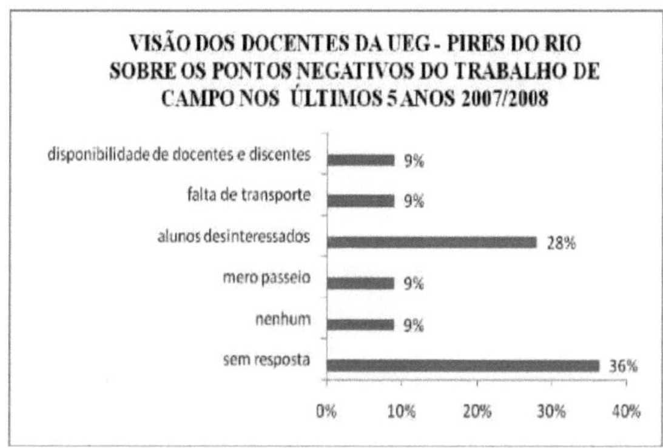

Graph 73: UEG - Pires do Rio teachers' views on the negative aspects of fieldwork in the last 5 years
Source: Research carried out from November 2007 to May 2008
Org.: CARNEIRO, V. A. - July/2008

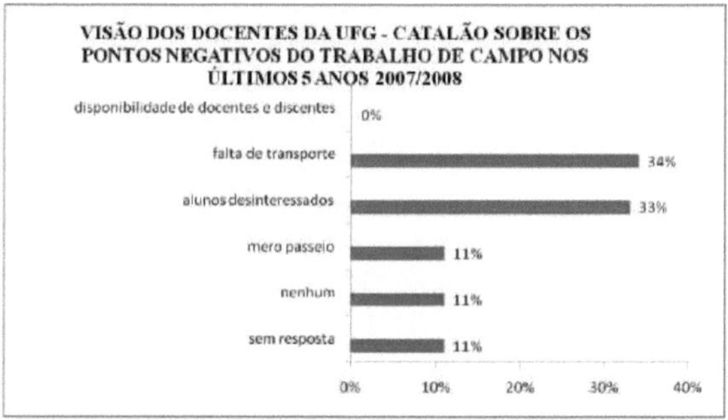

Graph 74: UFG - Catalão lecturers' views on the negative aspects of fieldwork in the last 5 years
Source: Research carried out from November 2007 to May 2008
Org.: CARNEIRO, V. A. - July/2008

Looking at graphs 75 and 76, we see a difference in the type of fieldwork between the teachers at the two institutions. According to the teachers at Pires do Rio (55%), the type of fieldwork that motivates them has more to do with the Geography course. In Catalão, 45% of the lecturers opted for trainer-type fieldwork, as we believe that it expands the students' possibilities in the labour market.

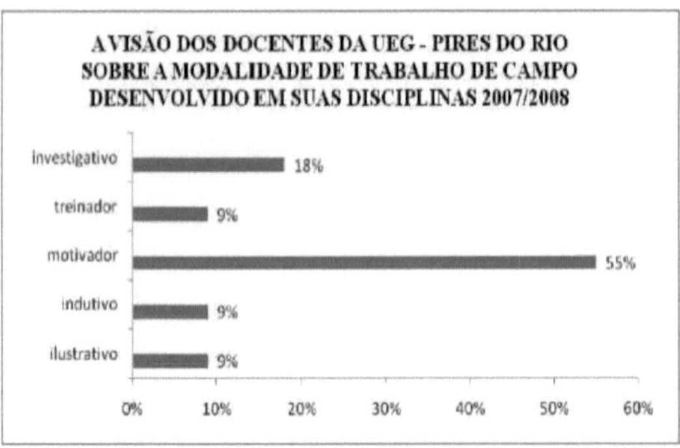

Graph 75: UEG - Pires do Rio lecturers' views on the type of fieldwork carried out in their subjects
Source: Research carried out from November 2007 to May 2008
Org.: CARNEIRO, V. A. - July/2008

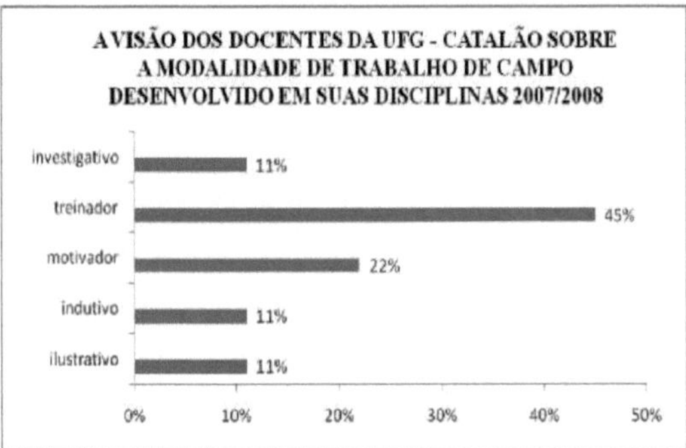

Graph 76: UFG - Catalão lecturers' views on the type of fieldwork carried out in their subjects
Source: Research carried out from November 2007 to May 2008
Org.: CARNEIRO, V. A. - July/2008

In Pires do Rio, there was a technical tie between the teachers on the issue of publicising field activities in geographic magazines and events. We found that 45 per cent of teachers support the publicity initiative and the other 46 per cent do not. We therefore believe that this mechanism is not yet mature enough for this to happen or that they are unaware of the significance of the experience report section in both journals and geographic events (Graph 77).

Graph 78 expresses the negative opinion of teachers in Catalão (78%) with regard to publicising their field activities in geography journals and events. We therefore come to the conclusion that the lack of interest in publicising and the lack of knowledge on the part of teachers with regard to experience reports are alarming and very worrying.

For Catalan teachers:

> "Students don't produce quality material for publication and dissemination" (Lecturer 6 - Catalão).
> "Our work is part of the syllabus of the subjects and not isolated projects aimed at literary production" (Teacher 8 - Catalão).

In Pires do Rio, the teachers are uncertain about the issue of publicising reports and articles in geography events and journals. So we collected a few statements:

"Usually at local events, because I think it's important to publicise the articles produced by the students" (Lecturer 10 - Pires do Rio).

"Sometimes. But a priori, the tendency, over the years, is to implement the preparation of scientific articles based on fieldwork" (Teacher 11 - Pires do Rio).

"It's important to publicise geography events, because it allows teachers and students to exchange experiences and methodologies" (Teacher 3 - Pires do Rio).

"I don't usually publish the reports, because they are restricted to quantitative evaluation" (Teacher 5 - Pires do Rio).

"Yes. The field activity reports are evaluated after the fact. Those that achieve scientific rigour and the students agree are sent to events. Work that doesn't achieve scientific rigour is instructed to be redone" (Teacher 8 - Pires do Rio).

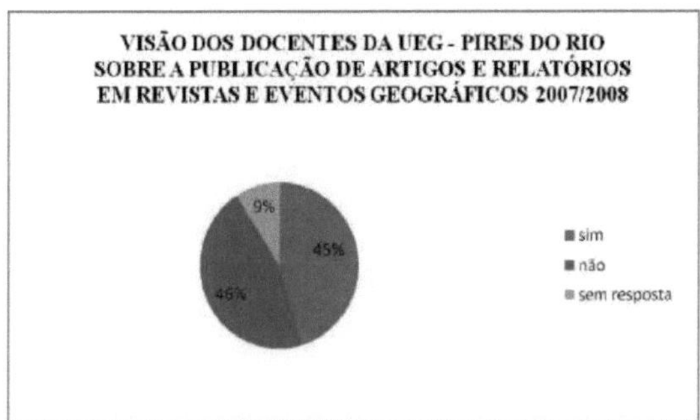

Graph 77: UEG - Pires do Rio lecturers' views on the publication of articles and reports in geographical journals and events

Source: Research carried out from November 2007 to May 2008

Org.: CARNEIRO, V. A. - July/2008

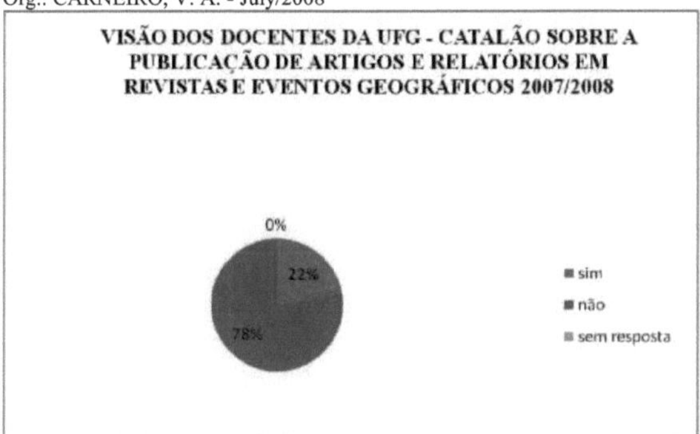

Graph 78: UFG - Catalão lecturers' views on the publication of articles and reports in geographical journals and events

Source: Research carried out from November 2007 to May 2008

Org.: CARNEIRO, V. A. - July/2008

Graphs 79 and 80 show that teachers from Pires do Rio (91 per cent) and Catalão (100 per cent) reported strong participation in field activities during their degrees. It's worth remembering that fieldwork is an important pedagogical activity and that beginners who are well connected to this activity gain a lot of experience for their professional future.

Let's look at some visions:

"Positive, because it broadened my knowledge and gave me experience that I could put into practice in high school and college" (Teacher 9 - Pires do Rio).

"Very positive, I had some good learning moments in the fieldwork. And most of the ones I took part in were just informative" (Teacher 1 - Pires do Rio).

"It was enriching because it allowed me to explore my field of work within the science of geography" (Teacher 3 - Pires do Rio).
"It made it possible to reconcile theory and practice" (Teacher 8 - Catalão).
"Excellent, it was even during field research that I came up with my master's topic" (Teacher 6 - Catalão).
"It was important for my training" (Teacher 1 - Catalão).

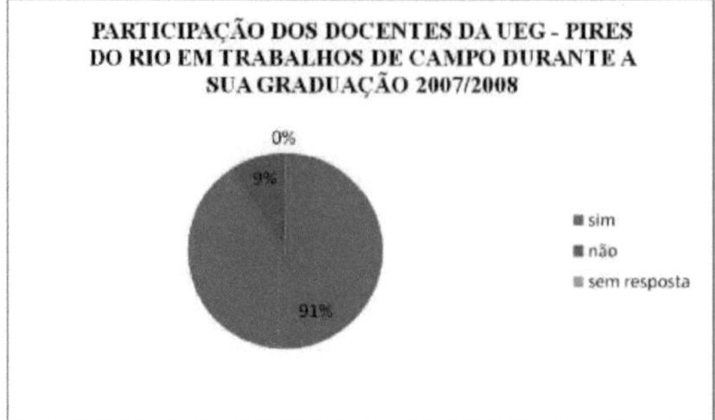

Graph 79: Participation of UEG - Pires do Rio lecturers in fieldwork during their undergraduate studies
Source: Research carried out from November 2007 to May 2008
Org.: CARNEIRO, V. A. - July/2008

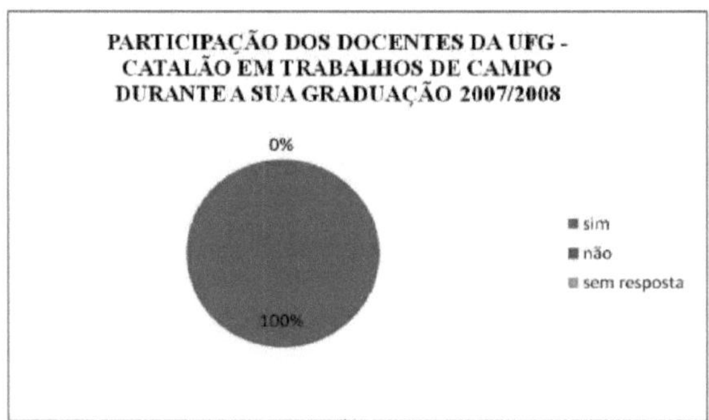

Graph 80: Participation of UEG - Pires do Rio lecturers in fieldwork during their degree programme
Source: Research carried out from November 2007 to May 2008
Org.: CARNEIRO, V. A. - July/2008

According to the teachers in Pires do Rio (82%) and Catalão (56%), students become interested in geographical subjects and/or themes for their monographs when they come across something striking in the fieldwork. In this case, we can highlight the fact that good planning, clearly covered content, technical mastery of the activity, the involvement and dynamic participation of the class, valuing the students' experiences and the didactic choice of a place to carry out the fieldwork are all factors that help students produce their monographs, either in the field of Human Geography or Physical Geography (graphs 81 and 82).

The teachers say:

"Yes. Several students choose their topic during the field research" (Teacher 6 - Catalão).
"It can help, because it enables interfaces between disciplines and researchers (teachers), as well as stimulating teamwork and the (theoretical-methodological) activities necessary for research" (Teacher 5 - Catalão).
"If fieldwork is motivating, it can arouse students' interest in a given problem or aspect

to be studied" (Teacher 6 - Pires do Rio).

"Practice relates to theory, in other words, students can be awakened by a problem they experience in fieldwork" (Teacher 5 - Pires do Rio).

"It allows students to see where they live and deal with issues in their locality, publicising their municipality" (Teacher 3 - Pires do Rio).

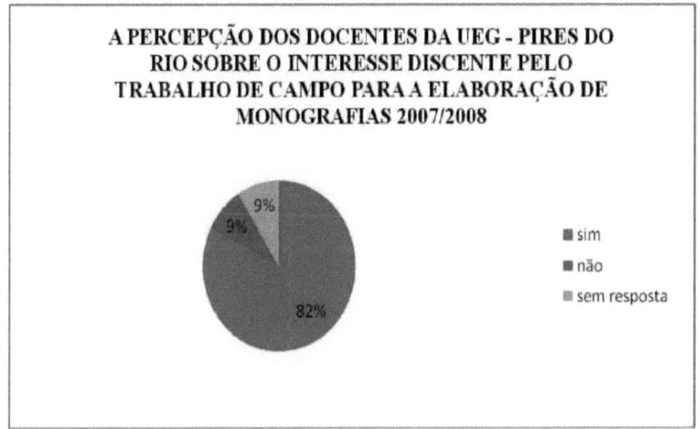

Graph 81: UEG - Pires do Rio lecturers' perception of student interest in fieldwork for monographs
Source: Research carried out from November 2007 to May 2008
Org.: CARNEIRO, V. A. - July/2008

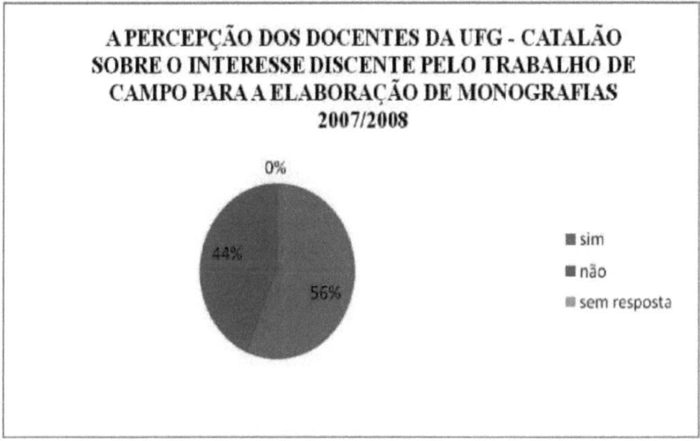

Graph 82: UEG - Pires do Rio lecturers' perception of student interest in fieldwork for monographs
Source: Research carried out from November 2007 to May 2008
Org.: CARNEIRO, V. A. - July/2008

Graphs 83 and 84 show that teachers in Pires do Rio (64%) and Catalão (67%) agree that fieldwork is an extremely important teaching resource for geographical knowledge.

Let's remember that "fieldwork is a well-known teaching method and at the same time a precious resource for pedagogical action, particularly in subjects such as Geography, History and Biology" (SILVA; PEDROSA, 2005, not paginated). Thus, according to Alegre (2006, p. 217), "doing Geography is a source of pleasure. Anyone who doesn't feel that pleasure should do something else".

In this way, we have

> "It's a teaching resource that combines the practical and the theoretical. It also gives students the chance to get involved in teaching and research by portraying their geographical space" (Teacher 3 - Pires do Rio).
> "It's a didactic activity outside the classroom with the aim of getting to know a particular geographical area" (Teacher 2 - Pires do Rio).
> "This didactic resource is very necessary, because Geography without fieldwork is not Geography" (Teacher 7 - Catalão).
> "Fieldwork is an indispensable teaching tool for the production of geographical knowledge" (Teacher 5 - Catalão).

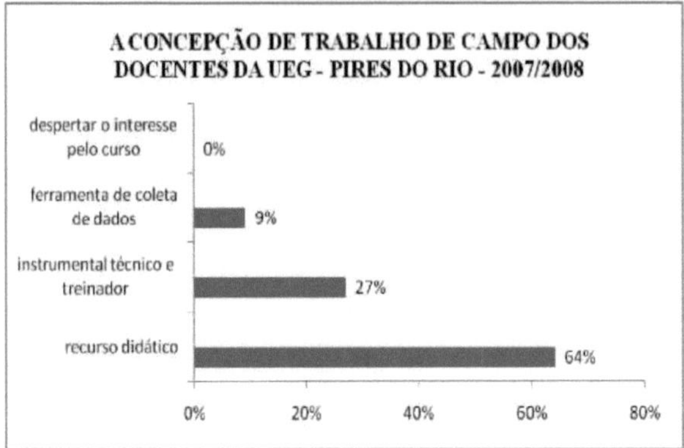

Graph 83: UEG - Pires do Rio teachers' concept of fieldwork
Source: Research carried out from November 2007 to May 2008
Org.: CARNEIRO, V. A. - July/2008

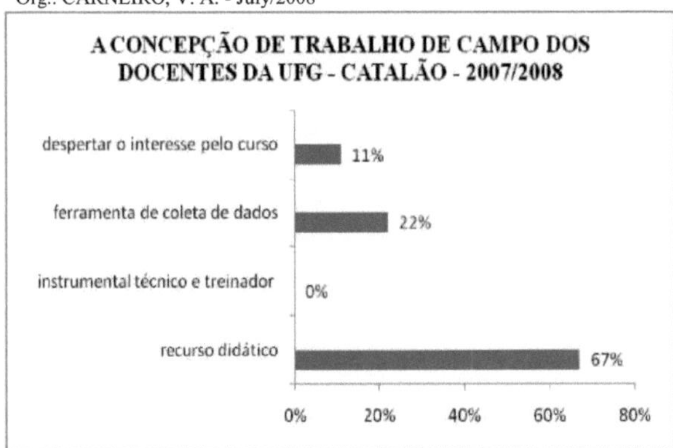

Graph 84: UFG - Catalão teachers' concept of fieldwork
Source: Research carried out from November 2007 to May 2008
Org.: CARNEIRO, V. A. - July/2008.

Finally, the verification found some "mismatches" and a lack of clarity in graphs 73 and 77 of the Pires do Rio teachers. We realised that the teachers' failure to position themselves precisely in response to the questions led to dubious interpretations. Therefore, for the purposes of analysis, we consider the situation in graph 73 to be "teacher relapse". We believe that the question about listing the negative points of fieldwork led to a 36 per cent omission on the part of the teachers, as they left this item unanswered.

Continuing in Graph 73, 28 per cent of teachers said that students' lack of interest was a negative aspect of the field activity.

Graph 77 shows a technical tie between yes and no on the question of the dissemination and/or publication of reports and articles on the field activities carried out by students. We can see that yes got 45% and no 46%. So, once again, there is an impasse because, according to a thorough check of the questionnaires answered by the Pires do Rio teachers, indecision is once again on the scene.

Let's look at an example of this indecision below:

> "No. Only the result of a piece of work was presented at an event. Most of the time the CT is only presented to the students themselves and there is no production of texts resulting from it for dissemination" (Teacher 1 - Pires do Rio) (my emphasis).

In light of the comments made by lecturer 1 - Pires do Rio, the question arises: What is an event? What is dissemination? What is publication? In this case, we don't need to go to the dictionary to find out what these words mean, because they are part of the geographers' context.

Analysing graphs 75 and 76, respectively from teachers in Pires do Rio and Catalão, we find different positions on the issue of defining fieldwork modalities. In Pires do Rio, 55% of the teachers thought that fieldwork was motivational and in Catalão, 45% of the teachers said that fieldwork was coaching.

We realise that the motivating nature of the fieldwork chosen by the Pires do Rio teachers lies in the fact that the Geography degree is the institution's "flagship" course. As for the situation in Catalão, the teachers' choice of the training aspect of fieldwork leads us to see the degree in Geography as a "stepping stone" to the bachelor's degree in Geography at this institution.Finally, we will move on to the final considerations of this master's research.

CHAPTER 4

FINAL CONSIDERATIONS

This research enabled the author to deepen his knowledge of the didactic resource fieldwork in the field of Geographical Science.

In light of the dissertation's findings, we can strongly affirm that fieldwork is a substantial didactic resource (graphs 83 and 84) for working integratively with the various subjects in the curricula of the Federal University of Goiás, Catalão Campus and the State University of Goiás, Pires do Rio University Unit.

We therefore emphasise that Geography plays an important role in shaping this knowledge produced by fieldwork, as it is a tool for understanding the relationship between man and nature.

We can also emphasise that geographical fieldwork is a modality that allows us to reflect on the place, discuss the articulation between theory and practice and underpin teaching and learning.

In this context, field classes in the areas of Physical Geography, Human Geography and Geography Teaching motivate and train students to solve questions that arise in class and can be solved on the spot. However, we would emphasise that the area of Physical Geography has stood out for its frequent use of fieldwork, which can be a route to maturing the researcher/teacher (graphs 27 and 28).

The practice of fieldwork means valuing the discovery of the place, where it is not carried out at random, but rather clarifies, fixes, sees, reviews and experiences the topics covered by the subjects in the classroom and fosters debates and interventions in the socio-environmental reality of our students.

This confirms the teachers' choice to emphasise fieldwork as an extremely important didactic resource for geographical studies (graphs 83 and 84).

We therefore emphasise the need to go out into the field in order to have contact with reality, because it is through the field that data is renewed, interpreted and reinterpreted in order to enhance the process of building geographical knowledge.

Fieldwork is the central theme of this study, which in general develops a theoretical reflection on the presence of this didactic resource in geography and in Geography degree courses.

The research therefore addresses fieldwork as a didactic resource, appropriate to the various fields of knowledge, which has great emphasis in Geography, including in the university environment.

Through the questionnaires, it was possible to identify the respondents' (teachers and students) conceptions of fieldwork, which end up being reflected in their pedagogical actions.

From the students' point of view, graphs 33, 34, 43 and 44 show that fieldwork is important for retaining and assimilating the geographical content covered in class.

We understand that the view put forward by the students is not consistent, because fieldwork cannot be considered as "an illustration of classroom teachings" (COMPIANI; CARNEIRO, 1993, p. 96).

Graphs 61 and 62 show that the importance of field activities for teachers lies in strengthening the link between theory and practice.

Thus, we found that the teachers and students from Pires do Rio opted for the motivating conception of fieldwork and the teachers and students from Catalão chose the training conception of fieldwork (graphs 45 and 46).

The motivational aspect of fieldwork seeks to sharpen and enhance students' interest in tackling a given problem or aspect to be addressed, while the training approach to fieldwork refers to the use of technical and theoretical apparatus so that students can understand and interact with the places visited.

We infer that the possibilities for teachers to carry out field activities are diverse, each with its own didactic and methodological characteristics. The various types of field activities presented in this dissertation have some distinct didactic characteristics.

In this way, we emphasise that it is up to the teacher - based on the characteristics found in their students and workplaces - to establish the type of fieldwork modality that best suits them, depending on the conditions offered.

Furthermore, it is worth emphasising that the fieldwork is aimed at sharpening the clinical eye and finding mechanisms to combat the doldrums of traditionalist practices.

Finally, the research opens up avenues for future reflections and discussions on the use of fieldwork in Geography degrees and we hope that this small contribution, in the form of a dissertation, can guide other research.

In this conclusion to the work, we leave you with a geographical provocation: VILIPENDING FIELD WORK MEANS PULLING THE TAPE AND SHOOTING YOUR OWN FOOT. Therefore, from the perspective of fieldwork, "Geography is done with the feet and the head" (Teacher 3 - Catalão). In this way, we can tell you that the research into fieldwork will not stop here, as a doctorate is the next target.

CHAPTER 5

REFERENCES

AB'SABER, A. N. Natural ecosystems. In: LEITE, J. L. Problemas - chave do meio ambiente. Salvador: SGM/IG/UFBA, 1995.

AB'SABER, A. N. The domains of nature in Brazil: landscape potential. São Paulo: Ateliê, 2003.

ADAS, M. Geografia da América: aspectos da geografia física e social. São Paulo: Moderna, 1982.

ALEGRE, M. The seventy years of the AGB (1934 - 2004). Terra Livre, São Paulo, v. 1, n. 26, p. 213-221, 2006 (statement).

ALHO, C. J. R.; MARTINS, E. S. From grain to grain, the cerrado loses space: cerrado - impacts of the occupation process. Brasília: WWF/PROCER, 1995.

ALVES, V. E. L. Fieldwork: a geographer's tool. GEOUSP, São Paulo: n. 2, p. 85-89, 1997.

AMORIM, M. E. Fieldwork as a teaching resource in geography in environmental conservation units - Itapuã State Park. 2006. 170 f. Dissertation (Master's in Geography) - Institute of Geosciences, Federal University of Rio Grande do Sul. Porto Alegre, 2006.

ANDRADE, M. C. Caminhos e descaminhos da geografia. Campinas: Papirus, 1989.

ANDRADE, M. C. Geografia, ciência da sociedade: uma introdução à análise do pensamento geográfico. São Paulo: Atlas, 1987.

ANDRADE, K. S. Toponymic atlas of indigenous origin in the state of Tocantins - Projeto Atito. 2006. 207 f. Thesis (Doctorate in Linguistics) - Faculty of Philosophy, Letters and Human Sciences, University of São Paulo, São Paulo, 2006.

ANDRADE, M. C. O pensamento geográfico e a realidade brasileira. Boletim Paulista de Geografia, São Paulo, n. 54, p. 5-28, 1977.

ANDRADE, M. C. Pierre Monbeig and geographical thought in Brazil. Boletim Paulista de Geografia, São Paulo, n. 72, p. 63-82, 1994.

ANDRADE, M. C. The trajectory and commitments of Brazilian geography. In: CARLOS, A. F. A. Geography in the classroom. São Paulo: Contexto, 1999.

ANDRADE, M. C. Uma geografia para o século XXI. Campinas: Papirus, 1999.

AQUINO, R. S. L. et al. História das sociedades americanas. Rio de Janeiro: Ao Livro Técnico, 1990.

ARBEX JR., J.; OLIC, N. B. Rumo ao centro-oeste: o Brasil em regiões. São Paulo: Moderna, 1996.

AZEVEDO, A. Alexander Von Humboldt, naturalist and geographer. Boletim Paulista de Geografia, São Paulo, n. 32, p. 54-72, 1959.

BALDANI, R. C. Field activities in environmental education: collective construction of methodological guidelines. 2006. 128 f. Dissertation (Master's in Science Education) - Faculty of Sciences, São Paulo State University. Bauru, 2006.

BALDINO, J. M. Ensino superior em Goiás em tempos de euforia: da desordem aparente à expansão ocorrida na década de 80. 1991. 180 f. Dissertation (Master's in Education) - Faculty of Education, Federal University of Goiás. Goiânia, 1991.

BALZAN, N. C. Study of the environment. In: CASTRO, A. D. et al. Didática na escola média: teoria e prática. São Paulo, Edibell, 1969.

BARADEL, C. B. Didactics: theoretical contributions and teachers' conceptions. 2007. 95 f. Monograph (Course Conclusion Paper in Pedagogy) - Faculty of Sciences, São Paulo State University. Bauru, 2007.

BARBOSA, A. S. Biogeographic system of the cerrado: elements for its characterisation. Goiânia: UCG/ITS, 1996.

BERMAN, M. Tudo que é sólido desmancha no ar: a aventura da modernidade. São Paulo: Cia da Letras, 1998.

BERTRAN, P. Uma introdução à história econômica do centro-oeste do Brasil. Brasília: CODEPLAN/UCG, 1988.

BOLETIM PAULISTA DE GEOGRAFIA. São Paulo: AGB, n. 84, 2006 (special volume: Fieldwork).

BOLETIM PAULISTA DE GEOGRAFIA. São Paulo: AGB, n. 81, 2004 (Historical Edition).

BOLETIM PAULISTA DE GEOGRAFIA. São Paulo: AGB, n. 88, 2008 (1978+30: AGB on the move).

BORGES, B. G. The awakening of the dormant. Goiânia: UFG, 1990.

BRASILIA (Federal District). CODEPLAN - Central Plateau Development Company. Study of the potential of the municipalities in the Geo-economic Region of Brasília: Pires do Rio - GO. Brasília, 1982, 20 p.

BRAUN, A. M. S. Breaking down the walls of the classroom: fieldwork as a language in geography teaching. 2005. 162 f. Dissertation (Master's in Geography) - Institute of Geosciences, Federal University of Rio Grande do Sul. Porto Alegre, 2005.

BRETAS, G. F. História da instrução pública em Goiás. Goiânia: UFG, 1991.

BROEK, J. O. M. Initiation to the study of geography. Rio de Janeiro: Zahar, 1972.

BUENO, F. S. Dicionário escolar da língua portuguesa. Rio de Janeiro: FAE, 1991.

CALAÇA, M. Transformations of the agrarian space in the cerrado: infrastructure and modernisation of agriculture. In: REGIONAL GEOGRAPHY MEETING, VII, 2001. Quirinópolis Proceedings... Quirinópolis: State University of Goiás, 2001. (CD-ROM).

CALLAI, H. C. The formation of the geography professional. Ijuí: EdUnijuí, 1999.

CALLAI, H. C. Studying place to understand the world. In: CASTROGIOVANNI, A. C. (Org.).

Teaching geography: practices and textualisations in everyday life. Porto Alegre: Mediação, 2000.

CALLAI, H. C. et al. The study of the municipality and the teaching of geography. Ijuí, EdUnijuí, 1998.

CALLAI, H. C. O estudo do município ou a geografia das séries iniciais. In: CASTROGIOVANNI, A. C. Geografia em sala de aula: práticas e reflexões. Porto Alegre: AGB/EdUniversidade, 1998.

CAMPOS, M. D. Catalão: historical and geographical study. Goiânia: Bandeirante, 1979.

CAMPOS, R. R. Brazilian geography (1930 - 1990): a science under construction. 1996. 147 f. Qualification Report (Master's in Education) - Faculty of Education, Pontifical Catholic University of Campinas. Campinas, 1996.

CAPEL, H. Filosofía y ciencia en la geografia contemporánea: una introducción a la geografia. Barcelona: Barcanova, 1981.

CARDOSO, L. P. C. The venerable Geographical Society of Rio de Janeiro: paths and initiatives in the institutionalisation of geographical knowledge in the first half of the 20th century. In: REGIONAL HISTORY MEETING, XII, 2006, Rio de Janeiro. Proceedings... Rio de Janeiro: ANPUH-RJ, 2006 (CD-ROM).

CARLOS, A. F. A. Brazilian geography today: some reflections. Terra Livre, São Paulo, v. 1, n. 18, 2002.

CARMELENGO, L. I.; TORRES, E. C. Teaching geography through fieldwork. In: ASARI, A. Y. et al. (Org.). Multiple geographies: teaching - research - reflection. Londrina: AGB, 2004.

CARNEIRO, V. A. Fieldwork in physical geography: Serra da Cachoeira do Maratá, Pires do Rio (GO). Trilhos, Pires do Rio, v. 4, n. 4, p. 127-137, 2006.

CARNEIRO, V. A.; DIAS, C. Field lessons in the gullies of Palmelo - GO. In: SYMPOSIUM ON TROPICAL SOILS AND EROSIVE PROCESSES IN THE WEST CENTRE, II, 2005, Goiânia. Proceedings... Goiânia: School of Civil Engineering / Federal University of Goiás, 2004 (CD-ROM).

CARNEIRO, V. A. An experience report of fieldwork on linear erosion processes in the peri-urban area of Palmelo - GO. Mediação, Pires do Rio, v. 2, n. 2, p. 35-44, 2007.

CARVALHO, D. A excursão geográfica. Revista Brasileira de Geografia, Rio de Janeiro, n. 4, p. 96-105, 1941.

CARVALHO, M. V. A. et al. In: SIMPÓSIO BRASILEIRO DE SENSORIAMENTO REMOTO, XIII, 2007. Florianópolis. Proceedings... Florianópolis: INPE, 2007 (CD-ROM).

CASTROGIOVANNI, A. C. Fieldwork in the teaching of geography in 1st and 2nd grade schools. Boletim Gaúcho de Geografia, Porto Alegre, n. 12, p. 71-74, 1984.

CASTROGIOVANNI, A. C.; SCHUTZ, L. S. The importance of fieldwork for teaching the humanities in primary and secondary schools. Cadernos do Aplicação. Porto Alegre, v. 1, n. 1, p. 43-48, 1986.

CAVALCANTI, L. S. Geografia e práticas de ensino. Goiânia: Alternativa, 2002.

CHAUL, N. F. Caminhos de Goiás: da construção da decadência aos limites da modernidade. Goiânia: UFG, 1997.

CHAUL, N. F. Catalan politics in the first republic. In: PALACÍN, L. et al. Political history of Catalão. Goiânia: UFG, 1994.

CHAVEIRO, E. F. The teaching of geography and the development of geographical thinking: elements for an evaluation of the UFG Geography Course. 1996. 167 f. Dissertation (Master's in Education) - Faculty of Education, Federal University of Goiás. Goiânia, 1996.

CHAVES, M. R. The legal exploitation of the Brazilian cerrado. In: REGIONAL GEOGRAPHY MEETING, VI, 1999. Catalão. Proceedings... Catalão: Federal University of Goiás, Catalão Campus, 1999. (CD-ROM).

CHAVES, M. R. The geography of the cerrado. Espaço em Revista, Catalão, v. 1, n. 1, p. 81-95, 1996.

CHAVES, M. R. et al. Insertion, criticism and intervention in reality: the AGB and Geography in Catalão - Goiás. Terra Livre, São Paulo, v. 1, n. 22, p. 133-143, 2004.

CHAVES, M. R. Decentralisation of environmental policy in Brazil and the management of natural resources in the cerrado of Goiás. 2003. 186 f. Thesis (Doctorate in Geography) - Institute of Geosciences and Exact Sciences, Paulista State University. Rio Claro, 2003.

CHAVES, M. R. The legal devastation of the cerrado and charcoal production in Catalão - GO. 1998. 139 f. Dissertation (Master's in Geography) - Faculty of Science and Technology, São Paulo State University. Presidente Prudente, 1998.

CHRISTOFOLETTI, A. The characteristics of the new geography. Geografia, Rio Claro, v. 1, n. 1, p. 3-33, 1976.

CHRISTOFOLETTI, A. Perspectivas da geografia. São Paulo: Difel, 1985.

CIDADE, L. C. F. Visões de mundo, visões da natureza e a formação de paradigmas geográficos. Terra Livre, São Paulo, n. 17, p. 99-118, 2001.

CLAVAL, P. The landscape of geographers. In: CORRÊA, R. L. e ROSENDAHL, Z. (Org.). Paisagens, textos e identidade. Rio de Janeiro: EdUERJ, 2004.

COELHO, M. A.; TERRA, L. Geografia geral: espaço natural e socioeconômico. São Paulo: Moderna, 2001.

COELHO, A. M. S. Towards a characterisation of geographical reasoning. 1997. 152 f. Dissertation (Master's in Geography) - Institute of Geosciences, Federal University of Minas Gerais. Belo Horizonte, 1997.

COLTRINARI, L. Fieldwork in 21st century geography. In: COLLOQUIUM GEOGRAPHICAL DISCURSE AT THE DAWN OF THE 21st CENTURY, 1996. Florianópolis. Proceedings... Florianópolis: Postgraduate Programme in Geography / Federal University of Santa Catarina, 1996. (CD-ROM).

COMPIANI, M. The relevance of field activities in the teaching of geology in the training of science teachers. Cadernos IG/UNICAMP. Campinas, v. 1, n. 2, p. 2-25, 1991.

COMPIANI, M. Doing geology with an emphasis on the field in the training of science teachers for 1st grade (5ª to 8ª grades). 1988. 246 f. Dissertation (Master's in Education) - Faculty of Education, State University of Campinas. Campinas, 1988.

COMPIANI, M.; CARNEIRO, C. D. R. The didactic roles of geological excursions. In: ENSENANZA DE LAS CIENCIAS DE LA TIERRA, 1993, [S. l.]. Proceedings... [S.l.]: ENSENANZA DE LAS CIENCIAS DE LA TIERRA, 1993. (CD-ROM).

CONTI, J. B. 1978+30: AGB on the move. Boletim Paulista de Geografia. São Paulo: AGB, n. 84, 2008 (statement).

CORRÊA, R. L. Fieldwork and globalisation. In: COLLOQUIUM THE GEOGRAPHICAL SPEECH AT THE DAWN OF THE 21st CENTURY, 1996. Florianópolis. Proceedings... Florianópolis: Postgraduate Programme in Geography / Federal University of Santa Catarina, 1996. (CD-ROM).

COSTA, F. A. G. Municipal socio-economic information for Pires do Rio. Goiânia: SEBRAE/GO, 1997.

COSTA, B. B. P. P. M. Fieldwork in the teaching of natural sciences: a study with teachers and textbooks from the 2nd cycle of basic education. 2006. 199 f. Dissertation (Master's in Education) - Institute of Education and Psychology, University of Minho. Minho (Portugal), 1996.

CRUZ, R. C. A. The paths of field research in geography. GEOUSP, São Paulo, n. 1, p. 93-98, 1997.

DAVID, C. Fieldwork: limits and contributions to geographical research. GeoUERJ, Rio de Janeiro, n. 11, p. 19-24, 2002.

DIAS, B. F. S. Alternativas de desenvolvimento dos cerrados: manejo e conservação dos recursos naturais renováveis. Brasília: IBAMA/FUNATURA, 1992.

DIAS, C. Mapping the municipality of Pires do Rio - GO: using geoprocessing techniques. 2008. 186 f. Dissertation (Master's in Geography) - Institute of Geography, Federal University of Uberlândia. Uberlândia, 2008.

DOLLFUS, O. O espaço geográfico. Rio de Janeiro: Bertrand Brasil, 1991.

DOURADO, L. F. A interiorização do ensino superior e a privatização do público. Goiânia: UFG, 2001.

DREW, D. Processos interativos homem-me ambiente. Rio de Janeiro: Bertrand Brasil, 1989.

ELIAS, D. Fieldwork: theoretical and methodological notes. GEOUSP, São Paulo, n. 5, p. 97-108, 1999.

EVANGELISTA, H. A. Geography at the Brazilian university. Revista Geo-Paisagem, Niterói, n. 6, 2004, 18 p.

EVANGELISTA, H. A. Geografia tradicional do Brasil: uma geografia tão mal afamada quanto mal conhecida. Revista Geo-Paisagem, Niterói, n. 19, 2006, 27 p.

EVANGELISTA, H. A. O XVIII Congresso Internacional da União Geográfica Internacional - UGI (Rio de Janeiro, 1956). Revista Geo-Paisagem, Niterói, n. 5, 2004, 28 p.

FANTINEL, L. M. Práticas de campo em geologia introdutória: papel das atividades de campo no ensino de fundamentos de geologia no curso de geografia, Universidade Federal de Minas gerais (UFMG). 2000. 144 f. Dissertation (Master's in Geosciences) - Institute of Geosciences, State University of Campinas, 2000.

FERNANDES, J. A. B. Do you see this adaptation? The science field lesson between the rhetorical and the empirical. 2007. 326 f. Thesis (Doctorate in Education) - Faculty of Education, University of São Paulo. São Paulo, 2007.

FERREIRA, A. M. Pires do Rio: the consolidation of a railway town. In: CHAUL, N. F.; SILVA, L. S. D. As cidades dos sonhos. Goiânia: UFG, 2005.

FERREIRA, I. M. Cerrado biome: environmental impacts and perspectives. In: REGIONAL GEOGRAPHY MEETING, VII, 2001. Quirinópolis Proceedings... Quirinópolis: State University of Goiás, 2001. (CD-ROM).

FERREIRA, I. M. Geomorphological models of veredas in cerrado environments. Espaço em Revista, Catalão, v. 7/8, n. 1, p. 7 - 16, 2005/2006.

FERREIRA, I. M. Paisagens cotidianas: veredas, a experiência dos moradores. Espaço em Revista, Catalão, v. 2, n. 2, p. 80 - 86, 1999.

FERREIRA, I. M. O afogar das veredas: uma análise comparativa espacial e temporal das veredas do chapadão de Catalão (GO). 2003. 242 f. Thesis (Doctorate in Geography) - Institute of Geosciences and Exact Sciences, Paulista State University. Rio Claro, 2003.

FERREIRA, C. L. A UEG no olho do furacão: o processo de criação, estruturação da Universidade Estadual de Goiás. 2006. 112 f. Dissertation (Master's in Education) - Faculty of Education, Federal University of Goiás. Goiânia, 2006.

FONSECA, F. P.; KUVASNEY, E. Multidisciplinary fieldwork: industries, settlements and conservation unit (Vassununga) along the Via Anhanguera - SP. GEOUSP, n. 13, 2003, 7 p.

FONTANA, J. Introduction to the study of general history. Bauru: EDUSC, 2000.

FORUM OF GOIAN NGOs FOR THE ENVIRONMENT. Cerrado Biome: subsidies for studies and actions, 1991. Goiânia: UCG/ITS, n. 1, 1991, 18 p.

FRANCO, J. M. V.; UZUNIAN, A. Brazilian Cerrado. São Paulo: Harbra, 2004.

FREITAS, I. N. A. Para pensar um novo mundo: a geografia dos jesuítas no Brasil. Mercator, Fortaleza, n. 3, 2003, 14 p.

GADOTTI, M. História das ideias pedagógicas. São Paulo: Ática, 1996.

GELPI, A.; SCHAFFER, N. O. Urban route guide. In: CASTRIGIOVANNI, A. C. et al (Org.). Geography in the classroom: practices and reflections. Porto Alegre: UFRGS/AGB, 1999.

GHIRALDELLI JR., P. História da educação. São Paulo: Cortez, 2000.

GIL, A. C. Como elaborar projetos de pesquisa. São Paulo: Atlas, 1999.

GOMES, H. A produção geográfica em Goiás. Goiânia: UFG, 1999.

GOMES, H.; TEIXEIRA NETO, A. Geografia: Goiás/Tocantins. Goiânia: UFG, 1993.

GOMES, M. A. F. et al. Agricultural use of the recharge areas of the Guarani Aquifer in Brazil and implications for groundwater quality. Jaguariúna: EMBRAPA - Meio Ambiente, 1999, 13 p.

GREGORY, K. J. The nature of physical geography. Rio de Janeiro: Bertrand Brasil, 1992.

GUERRA, A. T. Dicionário geológico - geomorfológico. Rio de Janeiro: IBGE, 1989.

GUERRA, M. O.; CASTRO, N. C. Como fazer um projeto de pesquisa. Juiz de Fora: EdUFJF, 1994.

GUIMARÃES, E. M. A. Fieldwork in river basins: the paths of an environmental education experiment. 1999 184 f. Dissertation (Master's in Geosciences) - Institute of Geosciences, State University of Campinas. Campinas, 1999.

HELFERICH, G. Humboldt's cosmos. Rio de Janeiro: Objetiva, 2005.

HISSA, C. E. V.; OLIVEIRA, J. R. O trabalho de campo: reflexões sobre a tradição geográfica. Boletim Goiano de Geografia, Goiânia, v. 24, n. 1-2, p. 31-41, 2004.

INOCÊNCIO, M. E. The PRODECER territory in south-eastern Goiás: Paineiras colonisation project. Mediação, Pires do Rio, v. 1, n. 1, p. 112-134, 2006.

JOHNSTON, R. J. Geography and geographers: Anglo-American human geography since 1945. São Paulo: Difel, 1986.

JUNQUEIRA, A. R. M. Fieldwork: an important didactic-pedagogical resource for secondary education - my experiences. 2003. 170 f. Dissertation (Master's in Geography) - Institute of Geography, Federal University of Uberlândia. Uberlândia, 2003.

KAYSER, B. The geographer and field research. Seleção de Textos, São Paulo, n. 11, p. 2540, 1985.

KOHLHEPP, G. Scientific discoveries of Alexander Von Humboldt's Expedition in Spanish America (1799-1804) from a geographical point of view. Journal of Biology and Earth Sciences. Rio de Janeiro, v. 6, n. 2, p. 260-278, 2006.

KRAEMER, E. Q. Teaching geography through fieldwork practices. 2003. 80 f. Monograph (Coursework in Geography) - Department of Social Sciences, Regional University of the Northwest of the State of Rio Grande do Sul. Ijuí, 2003.

KURY, L. Seeing with your own eyes: the traveller - naturalist. In: NATIONAL MEETING ON THE HISTORY OF GEOGRAPHICAL THINKING, I, 1999, Rio Claro. Proceedings... Rio Claro: Universidade Estadual Paulista / Instituto de Geociências e Ciências Exatas, 1999. (CD-ROM).

LACOSTE, Y. Research and fieldwork: a political problem for researchers, students and citizens. Seleção de Textos, São Paulo, n. 11, p. 1-23, 1985.

LEINZ, V.; AMARAL, S. E. Geologia geral. São Paulo: Nacional, 1989.

LEITE, M. L. M. Travelling naturalists. História, Ciências, Saúde - Manguinhos, Rio de Janeiro, v. 1, n. 2, p. 7-19, 1995.

LEME, R. B. Fieldwork: geography's laboratory. Visão Geográfica, Francisco Beltrão, v. 1, n. 1, p. 42, 1999.

LEPSCH, I. F. Formation and conservation of soils. São Paulo: Oficina de Textos, 2002.

LESTINGE, Environmental educators' views on environmental studies and belonging. 2004. 263 f Thesis (Doctorate in Forest Resources) - Luiz de Queiroz College of Agriculture, University of São Paulo, São Paulo, 2004.

LIMA, M. A. The golden years of Brazilian geography: background, achievements and consequences of the 50s and 60s. Revista Geo-Paisagem, Niterói, n. 3, 2003, 4 p.

LIMA, V. B.; ASSIS, L. F. Mapping some fieldwork itineraries in Sobral (CE): a contribution to geography teaching. Revista da Casa da Geografia de Sobral, Sobral, v. 6/7, n. 1, p. 109-121, 2004/2005.

MACHADO, M. S. Geography and epistemology: a walk through the concepts of space, territory and territoriality. GeoUERJ, Rio de Janeiro, n. 1, p. 17-32, 1997.

MAGALHÃES FILHO, F. B. B. História econômica. São Paulo: Saraiva, 1983.

MALYSZ, S. T. Study of the environment. In: PASSINI, E. Y. (Org.). Prática de ensino de geografia e estágio supervisionado. São Paulo: Contexto, 2007.

MAMIGONIAN, A. 1978+30: AGB on the move. Boletim Paulista de Geografia. São Paulo: AGB, n. 84, 2008 (statement).

MAPATSE, M. V. F. The excursion in the teaching/learning process of geography: subsidies for its realisation in the context of grade 10ª in Sofala Province - Mozambique. 2006. 162 f. Dissertation (Interinstitutional Master's Degree in Education) - Pontifical Catholic University of São Paulo / Pedagogical University of Mozambique. São Paulo, 2006.

MARANDOLA JR., E.; LIMA, A. Fieldwork and landscape: multidimensionality and methodological possibilities. Ciência Geográfica, Bauru, v. 9, n. 2, 2003.

MARTINS, S. F. Notas sobre os limites e as possibilidades da contribuição do trabalho de campo para a pesquisa geográfica. GeoSul, Florianópolis, v. 17, n. 34, p. 137-146, 2002.

MATHEUS, E. H. C. What's behind a pan? A field activity as a path to a geographical perspective. In: REGO, N. et al. Geografia, Porto Alegre: Artmed, 2007.

MATHEUS, E. H. C. Possibilities and limitations of field activities as a strategy for teaching geography. 2005. 152 f. Dissertation (Master's in Geography) - Institute of Geosciences, Federal University of Rio Grande do Sul. Porto Alegre, 2005.

MATOS, P. F. The consolidation of agricultural modernisation in the cerrado and its environmental impacts. Mediação, Pires do Rio, v. 1, n. 1, p. 66-81, 2006.

MELO, A. A. Trajectories of geography teaching in Brazil: 1978 - 1996. 2001. 176 f. Dissertation (Master's in Geography) - Institute of Geography, Federal University of Uberlândia. Uberlândia, 2001.

MENDES, E. P. P. Sugestões para elaboração de projetos de pesquisa: questões conceptuais e metodológicas. Catalão: CAC/UFG-Geography Course, 2005.

MENDONÇA, F. A. Geography and the environment. São Paulo: Contexto/UEL, 1993.

MENDONÇA, F. A. Geografia socioambiental. Terra Livre, São Paulo, n. 16, p. 113-132, 2001.

MENDONÇA, M. R.; THOMAZ JÚNIOR, A. The modernisation of agriculture in the cerrado areas of Goiás (Brazil) and the impacts on labour. Investigaciones Geográficas, Ciudad de México, n.55, p. 97-121, 2004.

MENEZES, C. O que é ser geógrafo: memórias profissionais de Aziz Nacib Ab'Saber. Rio de Janeiro: Record, 2007.

MESQUITA, H. A. Hydroelectric dams: the latest threat to the cerrado biome. Revista da Universidade Federal de Goiás, Goiânia, n. 1, p. 2124, 2005.

MIRANDA, L. F. A. The backlands of travellers. In: REGIONAL HISTORY MEETING, XIX, 2008, São Paulo. Proceedings... São Paulo: ANPUH/SP-University of São Paulo, 2008. (CD-ROM).

MOÇO, A. The world inside and outside school. Revista Nova Escola, São Paulo, n. 217, p. 7175, 2008.

MONTEIRO, C. A. F. A geografia do Brasil (1934 - 1977): avaliação e tendências. São Paulo: IG/USP, 1980.

MONTEIRO, C. A. F. Geography in Brazil throughout the 20th century: a panorama. São Paulo: AGB/Borrador, 2002.

MONZÓN, C. A. B. A geohistorical view of Venezuela in the work of Alejandro Von Humboldt. Geoensenanza, Mérida, v. 8, n. 2, p. 41-51, 2003.

MORAES, A. C. R. Geografia: pequena história crítica. São Paulo: Hucitec, 1998.

MOREIRA, J. R. Practical work in science learning - an innovative perspective: from theoretical foundations to practical material construction. In: NATIONAL MEETING ON SCIENCE EDUCATION / MEETING ON EDUCATION FOR A NEW WATER CULTURE, XI / 1st, 2005. Porto (Portugal). Proceedings... Porto: School of Education / Polytechnic Institute of Porto, 2005 (CD-ROM).

MOREIRA, R. Ten years have passed (the renewal of geography in Brazil: 1978 - 1988). Caderno Prudentino de Geografia, Presidente Prudente, n. 14, p. 5-39, 1992.

MOREIRA, R. What is geography? São Paulo: Brasiliense, 1994.

MOREIRA, R. Thinking and being in geography. São Paulo: Contexto, 2007.

MULLER, N. L. Aspectos da vida da Associação dos Geógrafos Brasileiros. Boletim Paulista de Geografia, São Paulo, n. 38, p. 43-73, 1961.

MULLER, N. L. Carl Ritter, the man and the geographer. Boletim Paulista de Geografia, São Paulo, n. 33, p. 78-88, 1959.

NASCIMENTO, I. V. Cerrado: fire as an ecological agent. Goiânia: UCG/ITS, 2001.

NASCIMENTO, M. A. L. S.; CASSETI, V. Geografia física do cerrado e impactos decorrentes do processo de ocupação. Goiânia: IESA-UFG, 1999.

NISKIER, A. Educação brasileira: 500 anos de história, 1500 - 2000. São Paulo: Melhoramentos, 1989.

NOGUEIRA, W. C. Pires do Rio: a landmark in the history of Goiás. Goiânia: Roriz, 1977.

OLIVEIRA, J. R. Fieldwork and the teaching of geography. 2005. 144 f. Dissertation (Master's in Geography) - Institute of Geosciences, Federal University of Minas Gerais. Belo Horizonte, 2005.

OLIVEIRA, L. A situação da geografia entre as ciências. Geografia, Rio Claro, v. 1, n. 1, p. 53-61, 1976.

OLIVEIRA, J. F. The restructuring of higher education in Brazil and the process of metamorphosis of federal universities: the case of the Federal University of Goiás (UFG). 2000. 210 f. Thesis (Doctorate in Education) - Faculty of Education, University of São Paulo. São Paulo, 2000.

OLIVEIRA, A. U. 1978+30: AGB on the move. Boletim Paulista de Geografia. São Paulo: AGB, n. 84, 2008 (statement).

OTTOBELLI, D. Agricultural modernisation and the socio-spatial transformations of Caldas Novas/ GO. 2005. 131 f. Dissertation (Master's in Geography) - Institute of Geography, Federal University of Uberlândia. Uberlândia, 2005.

PAES, I. D. R. Pires do Rio: our land, our people. Goiânia: Kelps, 1991.

PAIXÃO, A. M. Field activities as strategies for meaningful learning in environmental education. 2005. 1°1 f. Dissertation (Master's in Science and Maths Teaching) - Postgraduate Directorate, Lutheran University of Brazil. Canoas, 2005.

PALACÍN, L. History of the city of Catalão. In: PALACÍN, L. et al. Political history of Catalão. Goiânia: UFG, 1994.

PALACÍN, L.; MORAES, M. A. S. History of Goiás (1722 - 1972). Goiânia: UCG, 1994.

PAZERA JR., E. The French and Anglo-Saxon contribution to the formation of Brazilian geographical thought. Boletim de Geografia, Maringá, n. 1, p. 33-36. 1988.

PEDONE, C. Fieldwork and qualitative methods: the need for new reflections from Latin American geographies. Scripta Nova, Barcelona, n. 57, 2000. 29 p.

PEIXINHO, D. M. The recent occupation of the cerrados. In: REGIONAL GEOGRAPHY MEETING, VII, 2001. Quirinópolis Proceedings... Quirinópolis: State University of Goiás, 2001. (CD-ROM).

PEREIRA, J. V. C. Reflexões à margem de quatro excursões geográficas (Goiânia, Cabo Frio, Itatiaia e Valedo Rio Doce). Boletim Geográfico, Rio de Janeiro, n. 5, p. 7-14, 1943.

PEREIRA, Q. E. et al. Importance of the practice of soil analysis in the field in the teaching of physical geography - case study: geography degree course at the State University of Feira de Santana - BA. In: BRAZILIAN SYMPOSIUM OF APPLIED PHYSICAL GEOGRAPHY, X, 2003, Rio de Janeiro. Proceedings... Rio de Janeiro: State University of Rio de Janeiro, 2003 (CD-

ROM).

PEREIRA, R. M. F. A. Da geografia que se ensina à gênese da geografia moderna. Florianópolis, UFSC: 1989.

PEREIRA, S. N. Obsessões geográficas: viagens, conflitos e saberes no âmbito da Sociedade de Geografia do Rio de Janeiro. Revista da SBHC, Rio de Janeiro, v. 3, n. 2, p. 112-124, 2005.

PERINI, T. I. The process of inclusion in higher education in Goiás: the vision of the excluded. 2006. 121 f. Dissertation (Master's in Education) - Faculty of Education, Catholic University of Goiás. Goiânia, 2006.

PESSÔA, V. L. S. Fundamentos de metodologia científica para elaboração de trabalhos acadêmicos: material para fins didáticos. Uberlândia: [s.n.], 2007.

PILETTI, N.; PILETTI, C. History of education. São Paulo: Ática, 2003.

PINHEIRO, A. C. O ensino de geografia no Brasil: catálogo de dissertações e teses. Goiânia: Vieira, 2005.

PINTO, M. L. C. Fieldwork and the learning process in search of a method. School Spaces. Ijuí, n. 47, p. 15-20, 2003.

PINTO, M. N. Cerrado: characterisation, occupation and perspectives. Brasília: UnB/SEMATEC, 1993.

PIRES DO RIO CITY HALL. Pires do Rio: building a city for everyone, 1997. Pires do Rio, 1997, 32 p.

MUNICIPALITY OF SÃO PAULO. Environmental studies and other activities for evening classes, 1992. São Paulo, 1992, 100 p. (Caderno 4 - Cadernos de Formação).

PRUNES, M. L. Study plan for a geographical excursion. Geographical Bulletin. Rio de Janeiro, n. 5, p. 63-67, 1943.

RAMOS, C. Catalan yesterday and today. Catalão: Kalil, 1984.

READ, H. H. Geology: an introduction to the history of the Earth. Lisbon: Europa-América, 1966.

RIBEIRO, M. L. S. História da educação brasileira: a organização escolar. Campinas: Autores Associados, 1993.

ROCHA, E. A. V. Evaluation of the evolutionary process and erosion dynamics: a case study in the municipality of Ipameri - GO. 2007. 110 f. Dissertation (Master's in Geography) - Institute of Geography, Federal University of Uberlândia, 2007.

ROCHA, P. A. S. P. Fieldwork in the process of citizens' scientific literacy: research carried out at Praia de Lavradores, Vila Nova de Gaia. 2003. 115 f. Dissertation (Master's Degree in Geology for Teaching) - Faculty of Sciences/Department of Geology, University of Porto. Porto (Portugal), 2003.

RODRIGUES, A. B.; OTAVIANO, C. A. Guia metodológico de trabalho de campo em geografia. Geografia, Londrina, v. 10, n. 1, p. 35-43, 2001.

RODRIGUES, A. J. Geografia: introdução à ciência geográfica. São Paulo: Avercamp, 2008.

ROMANELLI, O. O. História da educação no Brasil (1930 / 1973). Petrópolis: Vozes, 1986.

RUCKERT, A. A. O papel social e político da geografia no Brasil: subsídios à história do pensamento geográfico no Rio Grande do Sul. Boletim Gaúcho de Geografia, Porto Alegre, v. 22, p. 17-26, 1997.

RUELLAN, F. O trabalho de campo nas pesquisas originais de geografia regional. Revista Brasileira de Geografia, Rio de Janeiro, n. 1, p. 35-50, 1944.

SANSOLO, D. G. Fieldwork and the teaching of geography. GEOUSP, São Paulo, n. 7, p. 135-145, 2000.

SANSOLO, D. G. The importance of fieldwork in geography teaching and environmental education. 1996. 302 f. Dissertation (Master's in Geography) - Faculty of Philosophy, Letters and Human Sciences, University of São Paulo. São Paulo, 1996.

SANTANNA NETO, J. L. History of climatology in Brazil: genesis and paradigms of climate as a geographical phenomenon. Florianópolis: DG/CFH/UFSC, 2004.

SANTOS, R. J. Empirical research and fieldwork: some questions about geographical knowledge. Sociedade & Natureza, Uberlândia, v. 11, n. 21/22, p. 111-125, 1999.

SANTOS, M. Homeland of mediocrity. Revista Educação. São Paulo, [s.n.], p. 5-7, 2000.

SANTOS, D. 1978+30: AGB on the move. Boletim Paulista de Geografia. São Paulo: AGB, n. 84, 2008 (statement).

SELECTION OF TEXTS. São Paulo: AGB, n. 11, 1985.

SILVA, M. J. The historical construction of the Catalão Advanced Campus of the Federal University of Goiás: 1983 - 2002. 2005. 259 f. Dissertation (Master's in Education) - Faculty of Education, Federal University of Goiás. Goiânia, 2005.

SCORTEGAGNA, A. Fieldwork in introductory geology courses: Geography courses in the state of Paraná. 2001. 150 f. Dissertation (Master's Degree in Geosciences) - Institute of Geosciences, State University of Campinas. Campinas, 2001.

SCORTEGAGNA, A.; NEGRÃO, O. B. M. Field work in introductory geology: the autonomous outing and its didactic role. Terrae Didática, Campinas, v. n. 1, p. 36-43, 2005.

SILVA, A. C. Natureza do trabalho de campo em geografia humana e suas limitações. Journal of the Department of Geography. São Paulo, n. 1, p. 49-54, 1982.

SILVA, A. M. R. Fieldwork: the "walking" practice of doing geography. GeoUERJ, Rio de Janeiro, n. 11, p. 61-74, 2002.

SILVA, A. M. R.; FIOREZE, Z. G. In: ENCONTRO NACIONAL DE HISTÓRIA DO PENSAMENTO GEOGRÁFICO, I, 1999, Rio Claro. Proceedings... Rio Claro: Universidade Estadual Paulista / Instituto de Geociências e Ciências Exatas, 1999. (CD-ROM).

SILVA, C. M. A little piece of Goiás territory. Mediação, Pires do Rio, v. 1, n. 1, p. 93111, 2006.

SILVA, E. M.; MELO, A. A. Preliminary study on fieldwork from the perspective of inclusive education: a challenge for future geography teachers. Caminhos de Geografia, Uberlândia, v. 9, n. 25, p. 87-95, 2008

SILVA, R.; PEDROSA, L. E. Fieldwork as a didactic resource: scripts and methodologies for the urban space of Catalão. In: ENCONTRO REGIONAL DE GEOGRAFIA, IX, 2005, Catalão. Proceedings... Catalão: Federal University of Goiás, 2005 (CD-ROM).

SILVA, V. P. On the trails of research: the most important thing is to know "why?". Caminhos de Geografia. Uberlândia, v.5, n. 17, p. 48-53, 2006.

SILVA, Y. F. O. Teacher training in the context of inclusive education: a study of the State University of Goiás. 2005. 130 f. Dissertation (Master's in Education) - Faculty of Education, Federal University of Goiás. Goiânia, 2005.

SILVEIRA, M. R. Area of influence of the municipality of Pires do Rio: the "Trilho das Penas" Region. 2007. 154 f. Dissertation (Master's in Geography) - Institute of Socio-Environmental Studies, Federal University of Goiás. Goiânia, 2007.

SIQUEIRA, J. Um contrato singular e outros ensaios de história de Goiás. Goiânia: Kelps, 2006.

SODRÉ, N. W. Introduction to geography. Petrópolis: Vozes, 1976.

SOUZA, M. B. Geografia Física: balanço da sua produção em eventos científicos no Brasil. 2006. 337 f. Dissertation (Master's in Geography) - Faculty of Philosophy, Letters and Human Sciences, University of São Paulo. São Paulo, 2006.

SOUZA, M. V. M. Practical interdisciplinarity in geography: a model for integrated fieldwork between disciplines in the physical area of the UFU geography course. In: SIMPÓSIO REGIONAL DE GEOGRAFIA, II, 2003. Uberlândia. Proceedings... Uberlândia: Federal University of Uberlândia, 2003 (CD-ROM).

STERNBERG, H. O. Contribuição ao estudo da geografia. Rio de Janeiro: Ministry of Education and Health / Documentation Service, 1946.

SUERTEGARAY, D. M. A. Geography and fieldwork. In: COLLOQUIUM THE GEOGRAPHICAL SPEECH AT THE DAWN OF THE 21st CENTURY, 1996. Florianópolis. Proceedings... Florianópolis: Postgraduate Programme in Geography / Federal University of Santa Catarina, 1996. (CD-ROM).

SUERTEGARAY, D. M. A. Physical geography and geomorphology: a (re)reading. Ijuí, EdUNIJUÍ, 2002.

SUERTEGARAY, D. M. A. Field research in geography. Geographia Magazine. Niterói, v. 7, p.92-99, 2002

SZMRECSÁNYI, T. et al. Dimensions, risks and challenges of the current sugarcane expansion. Brasília: EMBRAPA, 2008.

TAVARES, J. M.; MARINHO, M. B. Institutional self-evaluation report. Pires do Rio: UEG, 2006.

TEIXEIRA, E. As três metodologias: acadêmica, da ciência e da pesquisa. Petrópolis: Vozes/Unama, 2006.

TOMAZ JÚNIOR, a. Geografia passo-a-passo (critical essays from the 90s). Presidente Prudente:

Centelha, 2005.

TOMITA, L. M. S. Fieldwork as a teaching tool in geography. Geografia. Londrina, v. 8, n. 1, p. 13-15, 1999.

TROPPMAIR, H. Biogeography and the environment. Rio Claro: Graff Set, 1989.

TROPPMAIR, H. Simple methodologies for researching the environment. Rio Claro: [s.n.], 1988.

STATE UNIVERSITY OF GOIÁS. Academic-pedagogical guidelines, 2007. Pires do Rio, 2007, 32 p.

STATE UNIVERSITY OF GOIÁS. Academic guide, 2006. Pires do Rio, 2006, 28 p.

STATE UNIVERSITY OF GOIÁS. Geography course pedagogical project, 2005. Anápolis, 2005, 117 p.

STATE UNIVERSITY OF GOIÁS. Pires do Rio University Regulations, 2002. Pires do Rio, 2002, 25 p.

STATE UNIVERSITY OF GOIÁS. Resolution - CsA - n° 04, 2001. Anápolis, 2001, 1 p.

STATE UNIVERSITY OF GOIÁS. General regulations, 2000. Anápolis, 2000, 43 p.

FEDERAL UNIVERSITY OF GOIÁS. Regulations of the Advanced Campus of Catalão, 1985. Catalão, 1985, 15 p.

FEDERAL UNIVERSITY OF GOIÁS. Political-Pedagogical Project for the Geography Course: degree and bachelor's programmes, 2005. Goiânia, 2005, 83 p.

VALVERDE, O. Prehistory of the AGB in Rio de Janeiro. Terra Livre, São Paulo, n. 10, p. 117-122, 1992.

VENÂNCIO, M. Territory of hope: territorial plots of family farming in the rural community of São Domingos in Catalão (GO). 2008. 183 f. Dissertation (Master's in Geography) - Institute of Geography, Federal University of Uberlândia. Uberlândia, 2008.

VERDESIO, J. J. The environmental perspectives of the Brazilian Cerrado. In: PINTO, M. N. Cerrado: characterisation, occupation and prospects. Brasília: UnB/SEMATEC, 1993.

VESENTINI, J. W. Para uma geografia crítica na escola. São Paulo: Ática, 1992.

VIEIRA, A. B.; PEDON, N. R. O papel das comunidades científicas: a AGB Nacional e a Seção Local de Presidente Prudente / SP. Terra Livre, São Paulo, v. 1, n. 22, p. 71-83, 2004.

VIVEIRO, A. A. Field activities in science teaching: investigating the conceptions and practices of a group of teachers. 2006. 174 f. Dissertation (Master's in Science Education) - Faculty of Sciences, São Paulo State University, Bauru, 2006.

WAIBEL, L. Chapters on tropical geography and Brazil. Rio de Janeiro: IBGE, 1979.

WOOLDRIDGE, S. W.; EAST, W. G. Espírito e propósitos da geografia. Rio de Janeiro: Zahar, 1967.

ZUSMAN, P. B. In search of the origins of AGB. Boletim Paulista de Geografia, São Paulo, n.

78, p. 7-32, 2001.

http://www.agb.org.br. Accessed on 10/07/2007.

http://www.agb.org.br. Accessed on 09/04/2008.

Interview with Aziz Nacib Ab'Saber. .Net, Brasília, July 1992. Available at: http://www.canalciencia.ibict.br. Accessed on 14/03/2007.

http://www.ueg.br. Accessed on 11/04/2007.

http://www.ueg.br. Accessed on 04/03/2008.

http://www.catalao.ufg.br/geografia. Accessed on 24/07/2007.

http://www.catalao.ufg.br/geografia. Accessed on 18/08/2008.

http://www.geografiauegpiresdorio.com.br. Accessed on 11/02/2007.

http://www.geografiauegpiresdorio.com.br. Accessed on 24/09/2008.

http://www.ufg.br. Accessed on 20/04/2007.

http://www.ufg.br. Accessed at on 30/07/2008.

CHAPTER 6

ANNEXES

FEDERAL UNIVERSITY OF GOIÁS INSTITUTE OF SOCIO-ENVIRONMENTAL STUDIES
POSTGRADUATE PROGRAMME IN GEOGRAPHY

Annex 1 - Teacher questionnaire

ATTENTION!
This questionnaire is part of my Master's research in Geography, which is being carried out under the supervision of Prof Dr Manoel Rodrigues Chaves. Its purpose is to collect data on the conceptions of fieldwork held by teachers at this Higher Education Institution.

REMARKS:

1. Your identity will be preserved.
2. Thank you very much for your co-operation.

TEACHER QUESTIONNAIRE

Identification

01. Sex: ( ) male ( ) female
02. Age: ( ) 20 to 29 years old ( ) 30 to 39 years old ( ) 40 to 49 years old ( ) over 50 years old
03. Place of work: ( ) UEG - Campus Pires do Rio ( ) UFG - Campus Catalão
04. Professional status: ( ) civil servant ( ) contracted / commissioned
05. Your place of birth (City / State):

Academic background

06. Did you have a degree? ( ) yes ( ) no
07. Did you do a bachelor's degree? ( ) yes ( ) no
08. Have you specialised? ( ) yes ( ) no
09. Have you completed a master's degree? ( ) yes ( ) no ( ) currently studying
10. Have you completed a doctorate? ( ) yes ( ) no ( ) in progress
11. Do you teach undergraduate Geography? ( ) yes ( ) no
12. How many years have you been a Geography teacher?
13. Do you use fieldwork in your subjects? ( ) yes ( ) no
14. What places did you visit together with the graduating class?
( ) In Goiás ( ) Outside Goiás:
15. How important do you think fieldwork is in a geography course?
16. What are the biggest difficulties you face when carrying out fieldwork?
17. Do you often carry out fieldwork? ( ) yes ( ) no
18. In what periods / dates?
( ) weekend ( ) holidays and recess ( ) July holidays ( ) normal school days
19. Do you have any research, teaching or extension projects on the subject of fieldwork?( ) yes ( ) no
20. Point out some aspects related to the fieldwork you have carried out in the last 5 years:
a) Positive:
b) Negatives:

21. How would you define the fieldwork carried out in your subjects?
( ) **illustrative**: it is traditional and reaffirms knowledge as a finished product.

( ) **Inductive**: aims to sequentially guide the processes of observation and interpretation so that students can solve a given problem.
( ) **Motivating**: it aims to arouse the students' interest in a given problem or aspect to be studied, valuing each person's experience and questions.
( ) **trainer**: aims to train skills, usually using apparatus, instruments or scientific apparatus.
( ) **Investigative**: this type of field trip allows students to solve certain problems in the field and the teacher acts as a guide.

22. Do you publish fieldwork reports in the form of articles in Geography magazines and events? ( ) yes ( ) no
23. During your degree, did you take part in fieldwork? ( ) yes ( ) no
24. In your opinion, does fieldwork encourage you to write your final year monograph?
( ) yes ( ) no
25. What is your concept of fieldwork?

FEDERAL UNIVERSITY OF GOIÁS INSTITUTE OF SOCIO-ENVIRONMENTAL STUDIES
POSTGRADUATE PROGRAMME IN GEOGRAPHY

Annex 2 - Student questionnaire

ATTENTION!
This questionnaire is part of my Master's research in Geography, which is being carried out under the supervision of Prof Dr Manoel Rodrigues Chaves, and its purpose is to collect data on the conceptions of fieldwork held by the students of this Higher Education Institution.

REMARKS:

3. Your identity will be preserved.
4. Thank you very much for your co-operation.

STUDENT QUESTIONNAIRE

Identification

01.Grade: ( ) 1st grade ( ) 2nd grade ( ) 3rd grade ( ) 4th grade
02. Sex: ( ) male ( ) female
03. Age: ( ) less than 20 years old ( ) 20 to 29 years old ( ) 30 to 39 years old
( ) from 40 to 49 years old ( ) over 50 years old
04. Place of study: ( ) Geography degree at UEG - Campus Pires do Rio
( ) Degree in Geography at UFG - Campus Catalão
05.Shift: ( ) morning ( ) afternoon ( ) evening
06. Your place of birth (City / State):
07. Where do you live? (City / State):
08. Do you work? ( ) yes ( ) no

Student training

08. Where did you go to primary school? ( ) Public school ( ) Private school
09. Where did you go to secondary school? ( ) Public school ( ) Private school
10. Why did you choose Geography?
11. What is geography to you?
12. What is the purpose of geography?
13. Does the content of Geography relate to your day-to-day life? ( ) yes ( ) no
14. Do you take part in fieldwork organised by the teachers? ( ) yes ( ) no
15. Which subject(s) have you studied so far that use fieldwork as a teaching resource?

16. Where does the fieldwork proposed by your institution's teachers take place?
( ) in Goiás ( ) outside Goiás
17. How important do you think fieldwork is in a geography course?
18. What difficulties do you face when carrying out fieldwork?
19. What are the best times to carry out fieldwork?
( ) weekends ( ) holidays and recess ( ) July holidays ( ) normal school days
20. Do you participate in research, teaching and extension projects on the subject of fieldwork?
( ) yes ( ) no
21. Point out some aspects related to the fieldwork you did during your geography course.
a) Positive:
b) Negatives:
22. How would you define the fieldwork carried out in the courses you've taken so far?
( ) **illustrative**: it is traditional and reaffirms knowledge as a finished product.
( ) **Inductive**: aims to sequentially guide the processes of observation and interpretation so that students can solve a given problem.
( ) **Motivating**: it aims to arouse the students' interest in a given problem or aspect to be studied, valuing each person's experience and questions.
( ) **trainer**: aims to train skills, usually using apparatus, instruments or scientific apparatus.
( ) **Investigative**: this type of field trip allows students to solve certain problems in the field and the teacher acts as a guide.
23. During primary and secondary school, did you take part in fieldwork?
( ) yes ( ) no
24. What is your concept of fieldwork?

Annex 3 - Letter to the State University of Goiás

Serviço Público Federal
Universidade Federal de Goiás
Campus Catalão

Catalão, 29 de outubro de 2007

Senhores Professores,

Em função de cumprimento de cronograma de pesquisa de Mestrado cuja temática versa sobre "Concepções sobre trabalho de campo em Geografia" desenvolvida pelo aluno Vandervilson Alves Carneiro junto ao IESA/UFG, necessitamos de colher a impressão de alunos e professores do curso de Geografia da Unidade de Pires do Rio. Para tanto, foi desenvolvido um questionário para ser aplicada entre os discentes e docentes.

Para o caso de pesquisa junto aos discentes o questionário, com a devida autorização da Coordenação do Curso, será entregue em sala de aula, na quarta-feira dia 07/11//2007 e recolhido no dia 08/11/2007.

No caso dos professores será deixado nos escaninhos no dia 07/11/2007 e recolhido no dia 29/11/2007.

Certos da atenção e do atendimento da comunidade acadêmica, antecipadamente agradecemos.

Prof. Vandervilson Alves Carneiro
Mestrando

Prof. Manoel Rodrigues Chaves
Orientador

08/11/2007

Annex 4 - Letter to the Federal University of Goiás

Serviço Público Federal
Universidade Federal de Goiás
Campus Catalão

Catalão, 29 de outubro de 2007

Senhores Professores,

Em função de cumprimento de cronograma de pesquisa de Mestrado cuja temática versa sobre "Concepções sobre trabalho de campo em Geografia" desenvolvida pelo aluno Vandervilson Alves Carneiro junto ao IESA/UFG, necessitamos de colher a impressão de alunos e professores do curso de Geografia do Campus de Catalão. Para tanto, foi desenvolvido um questionário para ser aplicada entre os discentes e docentes.

Para o caso de pesquisa junto aos discentes o questionário, com a devida autorização da Coordenação do Curso, será entregue em sala de aula, na terça-feira dia 30/10/2007 e recolhido no dia 01/11/2007.

No caso dos professores será deixado nos escaninhos no dia 30/10/2007 e recolhido no dia 26/11/2007, na segunda etapa da disciplina ministrada pelo Prof. Idelvone Mendes Ferreira.

Certos da atenção e do atendimento da comunidade acadêmica, antecipadamente agradecemos.

Prof. Vandervilson Alves Carneiro
Mestrando

Prof. Manoel Rodrigues Chaves
Orientador

Recebi em
Cat. 29/10/07

Prof. Dr. Idir de Paiva Bueno
Coordenador do Curso de Geografia/CAC/UFG

Printed by Books on Demand GmbH, Norderstedt / Germany